EXPERIMENT ELEVEN

EXPERIMENT ELEVEN

*Dark Secrets Behind
the Discovery of a
Wonder Drug*

PETER PRINGLE

Walker & Company
New York

Copyright © 2012 by Peter Pringle

Published by Walker Publishing Company Inc., New York
A Division of Bloomsbury Publishing

All papers used by Walker & Company are natural, recyclable products made from wood grown in well-managed forests. The manufacturing processes conform to the environmental regulations of the country of origin.

LIBRARY OF CONGRESS CATALOGING-IN-PUBLICATION DATA
HAS BEEN APPLIED FOR
ISBN: 978-0-8027-1774-0

Visit Walker & Company's Web site at www.walkerbooks.com

First U.S. edition 2012

1 3 5 7 9 10 8 6 4 2

Typeset by Westchester Book Group
Printed in the U.S.A. by Quad/Graphics, Fairfield, Pennsylvania

*To the researchers in science
who did the hard work, and never
reaped the glory.*

Complete honesty is of course imperative in scientific work.
—W. I. B. Beveridge, *The Art of Scientific Investigation,* 1957

Contents

PART IV: THE PRIZE

PART V: THE RESTORATION

PART I · The Discovery

1 · Zones of Antagonism

EVERY DAY AT DAWN IN THE summer of 1943, a young graduate student could be seen striding briskly across the peaceful campus of the College of Agriculture. He was short, wiry, and handsome, his sharp features focused intensely on his important mission. Even his clothes seemed an afterthought, a wrinkled white shirt and loose gray pants, worn and reworn like those of any devoted researcher surviving on a meager stipend and the excitement of his work. He came from the direction of the Plant Pathology greenhouse, where students were breeding new varieties for President Franklin Roosevelt's Victory Gardens. He hurried past the dairy, with its herd of experimental Holsteins, past the poultry house, where Rhode Island Reds competed with white Leghorns for egg-laying prowess, and finally arrived at the Georgian-style Administration Building, which celebrated the proud colonial history of Rutgers University.

Albert Schatz, the harried student, was the first to arrive each day at the Department of Soil Microbiology. He let himself into the empty building and descended quickly into the basement laboratory. He pulled on his long white lab coat made of heavy cotton, worn haphazardly like his clothes and with a tear down one side. Then he began work on his experiments, searching for new antibiotics among the microbes he had found in the farmyard soil.

On that morning of August 23, he sat at his workbench and opened his notebook. On page 32, in his meticulous cursive, he entered the date and the title of his new experiment, "Exp. 11 Antagonistic Actinomycetes."

3

Then underneath, with the precision of a ledger clerk, he wrote, "Control soils Nos. 2, 7A, 18A, leaf compost, straw compost and stable manure plated out on egg albumin agar. Transfer made from colonies of actinomycetes selected at random . . . by casual macroscopic observation." And he added, for accuracy, "Some actinomycetes obtained from plates of swabs of chickens' throats . . . from Miss Doris Jones."

It was the fourth summer of the world war and, although the New Jersey college farm was thousands of miles from the front lines, almost everything, even the plants in the greenhouse and the microbes in the soil, had some link to the war effort. Schatz had come to the college because he wanted to be a farmer, but now, aged twenty-three and at the start of his doctorate career, he found himself engaged in a special war-driven mission. Instead of hunting for microbes that could break down soil to make it more fertile, Schatz was part of a scientific race to find microbes capable of producing new and more powerful antibiotics. His Experiment 11 was a routine test, one of hundreds being made by other graduate students working on this nationwide project, but it would be much more than that.

BY 1943, ELEVEN U.S. drug companies were producing penicillin, the first antibiotic, discovered by Alexander Fleming in 1928. The first vials of penicillin were being rushed to the front to treat common infections from battle wounds. But in this war, a new threat had emerged: biological weapons. Allied intelligence reported that Germany and Japan would not hesitate to use bombs and shells filled with deadly germs like anthrax, cholera, typhoid, and even the plague. Penicillin had no effect on such diseases.

The researchers best suited to the task of finding new antibiotics were not at medical schools, but at the agricultural colleges. "Aggies" like Albert Schatz were quite familiar with microbes from the soil that were capable of producing a chemical toxin that killed off the harmful bacteria that might be used in biological weapons. Each time they grew these microbes in a petri dish, they saw the telltale clear zones, or "zones of antagonism," as they were known, killing fields measured in millimeters where one microbe battled another for space or food. If they dropped penicillin into a petri dish of typhoid bacteria or cholera or the plague, no clear zones would develop. For this job, a stronger antibiotic was needed, the kind that Albert Schatz was hoping to find.

The basement laboratory where Schatz worked was primitive and sparsely equipped, grim even by the utilitarian standards of the times. But to this determined young man it was a hideaway, a place where he could work uninterrupted. And he had volunteered for this isolation. His ambitious Ph.D. had two parts: One was to find antibiotics against cholera and typhoid; a second was to find a cure for the deadliest of all infectious diseases, tuberculosis. Over the previous two centuries, two billion people had died from tuberculosis, caused by a slow-growing, pickle-shaped microbe, *Mycobacterium tuberculosis*. It was highly contagious, spreading easily from a victim's sneeze or spit. Known as the "Great White Plague," TB cut down rich and poor alike, although in the Industrial Revolution it spread more easily in the crowded slums and factories. In wartime it spread quickly in bombed-out cities and crowded refugee camps. There was no effective cure. Doctors had tried everything from prolonged rest to fortify the body's resistance, to drastic collapsed-lung surgery, to a range of alchemies and the new so-called sulfa drugs discovered by German researchers from industrial dyes. Albert Schatz's professor, Dr. Selman Waksman, had warned of the risks. Waksman insisted that when Schatz handled the deadly TB microbe, he must work alone in the basement and never bring the germ to the labs on the third floor.

In recent days, Schatz's laboratory had taken on the appearance of an ancient dispensary. In one corner an aging autoclave, a kind of pressure cooker used for sterilizing glassware, hissed away. Conical glass flasks contained rich dark brews of meat extract similar to gravy, a concoction Schatz used to feed his microbes. On his bench were rows of petri dishes with microbe food in agar, the jellied extract of algae that microbiologists use for growing their molds. The microbes Schatz was cultivating came from the multicolored and somewhat mysterious group known as the actinomycetes, or "ray fungi." These are strange, thread-like creatures that first appeared more than four hundred million years ago. They have wispy hyphae like the tentacles of a jellyfish, half bacteria and half fungus, a sort of evolutionary link between the two. They were favorites of Dr. Waksman, and common in the farmyard soil, and they had already shown promise in producing antibiotics. They form strikingly beautiful colonies of blues, reds, and grayish greens, and in the soil they are responsible for the pleasant odor of earth after a light rain.

Albert knew where to look for them. His favorite hunting grounds included the compost heaps of moldering leaves and twigs outside Plant Pathology, and the college stables, where he filled pots of fresh horse manure. The richest for his experiments, he knew, was the freshest—less than twelve hours old. Each gram of soil or compost Schatz collected was teeming with millions of different microbes, but always some actinomycetes. He diluted the soil with tap water and let drops of the mixture fall onto petri dishes containing microbial food in the jellied agar. Then he incubated the dishes and watched the actinomycetes grow. Within a few days, he had good colonies of mold in his petri dishes. Some of them were surrounded by the telltale clear zones, indicating that they might be antibiotic producers. He chose his likely candidates for their robust look and their widest zones of antagonism, like a gardener spotting a sturdy shoot, or a farmer selecting a high-yielding crop for breeding. Then he tested them for their action against known disease-producing bacteria from the same group as the typhoid and cholera germs.

By mid-September, he had selected two strains of a species of a gray-green actinomycete named *griseus*, Latin for "gray." One strain had come from heavily manured farmyard soil and he named it 18-16, for the sixteenth strain of the eighteenth soil sample. The other came from a colleague, Doris Jones, as he had noted in his lab notebook. He named her strain D-1, for Doris. Much quicker than he had dared to hope, Schatz had become convinced that he had discovered a new antibiotic, the first to be found in the Department of Soil Microbiology for several months, and everyone was excited, especially Dr. Waksman.

But no one could yet know whether his discovery would be useful as a medicine; if it was powerful enough to destroy typhoid and cholera it might also destroy human body cells. And it was only the first stage of Schatz's project; he still had to see if his new antibiotic would be effective against the toughest germ that causes TB. Schatz checked and rechecked Experiment 11, running the same tests over and over again until he was sure that he had not made a mistake. Each effort was carefully recorded in his lab notebook.

By the middle of October, he had confirmed that 18-16 and D-1 were indeed behaving like good producers of a new antibiotic. On October 19 at two o'clock in the afternoon, he placed a culture of his *Actinomyces griseus* in a test tube and sealed it for posterity by heating the end over a Bunsen burner and twisting the glass shut. That weekend, he wrapped the tube in cotton wool, put it in his pocket, and caught the train from the Rutgers University town of New Brunswick to Newark, then the bus to Passaic, where his parents lived in a working-class section of the textile town on the Passaic River. There, he showed the test tube to his father, Julius, and his uncle Joe and presented it to his mother, Rachel. She had not finished grade school and had no real idea what the test tube represented. He told her that he had found a new medicine that might eventually fight the infectious diseases, maybe even tuberculosis, that she had seen too often destroying the lives of her friends and neighbors. That she could understand.

2 · The Apprentice and His Master

THE SCHATZ FAMILY CAME FROM THE peasant class in the old Russia, and their entry into America is an immigrant story of the kind often told at the turn of the twentieth century. Albert's grandfather, Shlomo (Sam) Schatz, was a butcher, and his grandmother's family, the Tunicks, were known for their physical strength and much revered in the community for forming local vigilante committees to defend Jews during the pogroms. Sam himself was a strong man who once, legend has it, leaped on a bull that was running amok through the village and wrestled it to the ground. But Russia was a barren and hostile place, especially for Jews, and Sam left his village on the outskirts of Minsk in 1899 and immigrated to America, leaving his pregnant wife, Rose, and their five children with her father, Ephraim. He arrived at Ellis Island and moved in with a cousin on Manhattan's Lower East Side. It took him five years, working as a housepainter, to save enough money to bring his family, including Albert's father, Julius, to New York.

The family lived in a walkup, and soon after their arrival one child died of a weak heart. They moved into a Brooklyn tenement, and Sam and Rose had six more children, but the man who could wrestle a bull grew weak from heavy smoking and living in the putrid city air. When doctors told him he should leave for a life in the country, the Jewish community had just the answer.

Baron Maurice de Hirsch, a German-born Jewish banker, gave mortgages to immigrant Jews to enable them to build their own barns and homes. He also set up the small Woodbine Agricultural College in New

Jersey to produce "intelligent, practical farmers." With the help of Hirsch funds, Sam Schatz bought a dirt farm in Fitchville, Connecticut, joining other Jewish settlers in small communities across the state. On most of these farms the soil was poor, exhausted by Yankee farmers who had abandoned it to move west or, in some cases, for better jobs in the cities.

The Schatzes were the first Jewish family at Bird's Eye View Farm, a stone house, two wooden barns, and a manure pit built on a rise known as Cannonball Hill. The family scratched a living from a dozen milk cows and some chickens. They sold vegetables in the spring, and in the summer they took in boarders from the city. While the urban renters lived in the farmhouse and enjoyed the great outdoors, the Schatz family lived in tents. Julius joined the U.S. Army in World War One, and after he returned, he was delivering vegetables by horse cart to nearby Norwich one day when he met a pretty, dark-haired young woman named Rachel Martin who worked in a bakery. They soon married. Her parents were Jewish immigrants from Poland who had come to America via Britain.

On February 2, 1920, Albert Schatz was born in a Norwich hospital. The family stayed on the farm until he was three, when they moved to Passaic, New Jersey, where Julius's sister Rebecca and her husband, Abe, had a grocery store. They lived in a wooden three-story house with six apartments, three at the front and three at the back. Two girls were born, and the family moved back and forth from Passaic to the farm, wherever there was work. As soon as he was able, Albert helped out on the farm. He learned how to sharpen farm tools, milk cows, make butter and cheese, and drive the horse cart. When he was older, he shot groundhogs, mended his own clothes, and darned his socks. He attended the local one-room schoolhouse, which had one teacher and twenty students, grades one through eight. The building was twenty by twenty-five feet and had two entrances, one for boys and one for girls. Albert wanted to be a farmer, like his father and grandfather.

During the Great Depression the family lived mostly in Passaic. They joined other immigrants from Eastern Europe—Hungarians, Czechs, Poles, and Russians. Albert witnessed much poverty and sickness, people fighting for scraps on the garbage dumps and dying from infectious diseases, like pneumonia, diphtheria, and, of course, tuberculosis. It was a raw and sometimes violent period. One of the young boy's lasting memories was of the bloody police charges that ended the fourteen-month-long Passaic

*Albert Schatz, age twelve, with his mother, sisters Sheila and Elaine,
and his maternal grandmother on the Connecticut farm in 1932.
(Courtesy Vivian Schatz)*

textile workers' strike involving fifteen thousand workers, in 1926–27. The
police dispersed the strikers with horses and water cannons, and schools
were often closed. Despite the disruptions, Albert managed to stay in classes
and was a consistently promising student at Passaic High School.

In his junior year, in 1936, when he was sixteen, he contributed three
paragraphs to the school newspaper about his "life's ambition," to be a
farmer. He did not seek wealth "for I should not know what to do with it."
He wanted to "sweat by honest labor" and to "roam the open fields." He
wanted to chop wood until his muscles ached. "I want to LIVE."

Aged eighteen, Albert won a scholarship to the Rutgers' College of
Agriculture, the first in his family ever to attend an institution of higher
learning. In his second year, he was elected to Phi Beta Kappa, a rare
achievement for an "aggie." The head of the Department of Soil Microbiology, Dr. Waksman, was another Jewish immigrant of Russian descent. He
was always on the lookout for bright young graduates, and was happy to
accept Albert as a Ph.D. candidate.

SELMAN ABRAHAM WAKSMAN, the man behind the intense wartime hunt for
antibiotics at Rutgers, was no ordinary soil scientist. Like the Schatz family, Waksman had arrived in America at the beginning of the twentieth

century, but he came from a different social order and had achieved much in the New World.

He was born on July 8, 1888, according to the old Russian calendar, in the small market town of Novaya Priluka, in western Ukraine, two hundred miles from the regional capital of Kiev. He wrote in his memoir that it was "a mere dot in the boundless steppes," surrounded by chernozem, the fertile black earth on which wheat, rye, barley, and oats flourished, as long as the rains came. Without them, famine swept the land. The inhabitants of the small towns and villages of Western Ukraine were recently freed serfs who scratched a living from smallholdings, and Jewish artisans and tradesman who marketed the farm and forest products.

His life there was simple, but not uncomfortable. His father was the relatively well-off son of a coppersmith and had inherited property. His mother was the daughter of a successful businesswoman who ran a dry goods store, a "prominent merchant in the community." His mother had inherited the store, and together his parents were able to pay for Selman's private tutors.

Immediately after marriage, his father had been drafted, like all able-bodied men, into the czar's army for five years, leaving Selman's mother to carry on her business and fend for herself. When his father had returned from service, Selman had been born, but his father showed little interest in being with his son, most of the time living twenty miles away in the nearest large city, Vinnitsa, where he had inherited property. Selman was brought up by his mother, several aunts and maiden cousins, and his maternal grandmother, who had eight daughters. Selman was the son of the youngest daughter. Inevitably, he was spoiled.

His mother taught him to read and sent him to the local heder (private school) and then to private tutors. She also made sure that he studied the Bible and the Talmud. The young Selman quickly learned Hebrew and Russian literature, history, and geography. And he was frequently picked as the one to read a chapter from the Bible or deliver the blessing on the initiate at a bar mitzvah.

Jews and Ukrainians lived side by side in Novaya Priluka. The Waksmans lived in the wealthier part of town. His mother gave birth to a daughter when Selman was seven, but the daughter died less than two years later of diphtheria.

In the Waksman household there was usually money left over to help a needy niece or nephew, or the less fortunate on the town's poorer side. Encouraged by his mother, Selman gave free lessons in Hebrew and Russian, and later private lessons to the sons of the wealthier inhabitants and the richer peasants.

The first Russian uprising of 1904–05 did not affect little Novaya Priluka, but revolution was in the air. Selman's friends were divided on the future. One believed that socialism was inevitable, and another, the Zionists, looked for salvation in a new homeland in Palestine. Selman was uncommitted, with divided sympathies—on the fringe of the two groups. Instinctively, he favored the revolutionaries, but he disliked the fierce arguments over the form of a future government, should a revolution be successful. He was more interested in pursuing a higher education, but the way was blocked because he was a Jew. He could not enter the gymnasium or go on to university without passing a special competitive exam.

In 1908, he left with four friends for Odessa to be coached, at a price, for the crucial exam. He passed "with flying colors" and returned home a hero now set to attend university in Odessa. But suddenly he suffered a terrible blow. In the summer of 1909, his mother died of an intestinal blockage. During the seven days of mourning, he read and reread the Bible, "perhaps for the last time."

He returned to Odessa to find new political barriers. Candidates for the university had to have been born in Odessa or have spent the last twenty years there. Selman managed to bribe a government official to give him the necessary papers, but when his friends were refused admission, they all decided to leave Russia for good. He thought briefly of going to Switzerland, a destination favored by his father, but his cousins in Philadelphia, having heard of his mother's death, urged him to join them.

In October 1910, Selman and a group of five young people from Novaya Priluka, three men and two women, left by train for Bremen, and thence for America. They landed in Philadelphia on November 2.

BY THE BEGINNING of the twentieth century more than three million Russians had immigrated to the United States. Waksman, now aged twenty-two, went to work on his cousin's five-acre farm near Metuchen, New Jersey, thirty miles from New York City. He helped with the hens, learned how to make compost from stable manures, and planted vegetables in the

Selman Waksman as he was about to leave Russia in 1910. (Special Collections and University Archives, Rutgers University Libraries)

spring for local markets. His cousin was a great teacher, and at the end of his first year Waksman published an article in the *Rural New Yorker* titled "How I Raised a Flock of Chickens," for which he was paid his first ten dollars.

But his goal was college. He thought of becoming a doctor, and was accepted at Columbia University medical school. Another cousin, who was a dentist, offered to help with the fees, but Waksman did not want to be tied down by debt.

So he had to take what was available, and in those days the most accessible institutions were the land grant colleges. These were created by the Morrill Act of 1862, which gave states land grants to fund public agricultural and engineering colleges. One of the first such establishments was at Rutgers College, established originally as Queen's College in 1766 and still a small institution at the turn of the twentieth century.

Rutgers was only eight miles from Metuchen, and Waksman's farmer cousin suggested he should go and see Jacob Lipman, another Russian immigrant, who was then head of the Department of Bacteriology. By 1911, Lipman was an established figure in soil science, having made his reputation on studies of bacteria that make nitrogen available for crops.

Waksman was persuaded that a course in agriculture would satisfy his curiosity about the biochemistry of living organisms, plus he was awarded a full scholarship. Aged twenty-three, he found it hard, at first, to be among much younger boys of seventeen, who teased him for his clumsy English and dislike of sports. He also found the level of teaching poor. In his sophomore year, his chemistry professor was "an unimaginative bore," physics was "a great disappointment," he found the courses on American and English literature uninteresting, and he disliked Shakespeare. The French teacher was enthusiastic, but he felt he already had enough knowledge of foreign languages. The only courses that earned his approval were zoology and botany.

At the end of his second year, he yearned for independence and moved into a room in an old house on the college farm, paying for his accommodation by working in the college greenhouse and helping out in the laboratory. He bought cracked eggs from the Poultry Department at eleven cents a dozen.

Another and more important reason for striking out on his own was the arrival in New York from Novaya Priluka of a young woman named Deborah Mitnick. The daughter of a prosperous grain merchant, she was the sister of Waksman's best friend, Peisi, back in Ukraine, and after finishing grade school she had come to stay with her cousins, braving the voyage from Riga on her own in the middle of winter. She was good looking, bright, and energetic. In America, she quickly joined Peisi in New York—he had come to America with Waksman, in 1910. She worked in a sweatshop, became a member of the International Ladies' Garment Workers' Union, and took singing lessons. She was affectionately known as Bobili, Russian for young grandmother, a nickname given in the hope that she would reach a ripe old age. Waksman had been her tutor in Novaya Priluka, had always admired her, and planned to marry her.

In his studies, Waksman had at last found a subject that interested him: general bacteriology under Dr. Lipman. "I felt that I was finally under the tutelage of a master," he wrote. Waksman was the only student majoring in soil microbiology. For his senior thesis he listed the different groups of microbes—bacteria and fungi—but he was fascinated by the actinomycetes. He dug trenches on the college farm and mapped the different horizons of the microbes he found in the soil. He took samples from each layer, suspended them in water, put the microbes on petri dishes of nutrient agar, let them grow for a week, and then counted the different colonies that had developed.

The actinomycetes, hardly noticed in America, had been known for more than forty years in Europe, having first been described by German researchers as a microbe responsible for a disease in cattle known as "lumpy jaw," literally lumps on the animal's cheek containing a growth of the microbe. Russian researchers had also published papers on the actinomycetes, and Waksman had a distinct advantage over his colleagues because he could read German and Russian. He cataloged the different species and played an important role in their early classification into five genera, depending on a microscopic examination of the degree of branching of the cells, whether they produced spores in chains or singly on stalks, and whether they could live with or without oxygen. Thus, as Waksman would write forty years later, began his interest "in a group of microbes to which I was later to devote much of my time and which were to remain for the rest of my life as my major scientific interest."

He was elected to the scholastic fraternity Phi Beta Kappa, and on his graduation in 1915, Lipman offered him a job as a research assistant in soil microbiology and a stipend of fifty dollars a month, which in those days meant he could continue to live comfortably in his room in the farmhouse and study for his master's degree.

By the end of 1915, he had written his first academic paper on bacteria, actinomycetes, and fungi in the soil. He was twenty-seven. The paper was published in February 1916, the year he became an American citizen and the year he married Deborah "Bobili" Mitnick.

Selman Waksman married Deborah Mitnick, his childhood sweetheart, in 1916.
(Special Collections and Archives, Rutgers University Libraries)

"I had sent my roots into the soil in search of its microbiological population," he later wrote. "I was now on my way. I knew now exactly what I wanted and how to get it. The rest was merely to follow a plan. California was to prove whether I was on the right track."

The new couple moved to the University of California at Berkeley, where Waksman studied for two years for his doctorate on the enzymes produced by microbes, mostly the actinomycetes. During his last year he supplemented his income by working at Cutter Laboratories, a local commercial laboratory producing antitoxins and vaccines against bacterial infections. It was the start of a lifelong connection between his research as a microbiologist and industry.

3 · The Good Earth

IN THE SUMMER OF 1918, AS World War One was drawing to a close, Waksman returned to Rutgers to take up a new position as the farm college "Microbiologist." His somewhat mundane task was to continue the search for microorganisms that would produce more fertile soils, but he insisted on the rather grand title, with a capital *M*, as a mark of the importance he attached to the emerging science, and his own place in it.

The war had taken its toll, even on the quiet backwater of the New Jersey Agricultural Experiment Station. There were no graduate students and no lab assistants in the Department of Soil Microbiology. Waksman found the laboratory benches covered with dirty petri dishes, and the cultures of fungi and actinomycetes he had put into the culture collection before leaving for California were either dead or in need of prolonged resuscitation.

The sorry state of the laboratories reflected the scarcity of funds. His mentor, Dr. Lipman, could offer him only one day a week and fifteen hundred dollars a year, less than he had been getting as a part-time bacteriologist in the Cutter Laboratories in California. Personally, he had no funds in reserve, and he was forced to look for another part-time job in industry to supplement his income. It did not present a problem.

Before the war, America had been dependent on Germany for supplies of chemicals, laboratory glassware, and even scientific literature, but was gradually severing these ties and establishing its own infrastructure of scientific research. Waksman was now a bacteriologist and a biochemist, a good combination for employment amid the expansion of microbial research after the war. The brewery and food industries were studying yeasts and cheese-producing molds, public health officials were looking at new ways to use microbes in sewage disposal, and drug companies were beefing up their research into medicines to cure infections. In agriculture, researchers concentrated on identifying soil-enriching microbes to grow bigger and better crops.

This bustling activity was a natural progression of the work of the nineteenth-century European pioneers of bacteriology—Louis Pasteur, who first formulated the germ theory of disease, and the German bacteriologist Robert Koch, who discovered the TB microbe. On the eve of World War One, German researchers had found a new purpose for the dyes that had been used to identify bacterial cells under the microscope. In 1910, a young German chemist, Paul Ehrlich, and his Japanese assistant, Sahachiro Hata, found an arsenic-based dye that worked against the syphilis microbe. They named it salvarsan, and it became the first of the so-called magic bullets that would cure bacterial infections.

From among the many companies making offers, Waksman chose the Takamine Laboratory, in nearby Clifton, New Jersey, one of the more successful of the new companies producing antibacterial products, including salvarsan. Waksman's job was simple enough for a postgraduate biochemist—he had to test each batch of salvarsan for toxicity to human cells. He was paid good money for those days, forty-five hundred dollars a year, and the company was close enough to Rutgers for him to combine his work with a day a week at the college. The money even allowed him to move into Manhattan, where his wife, Bobili, could enjoy the music, theater, and culture missing in rural New Jersey.

Over the next two years, Waksman was exposed to much more than how to test a drug for toxicity. The Takamine Laboratory produced and marketed adrenaline, a natural product of animal adrenal glands. His experience "suggested the possibility" of finding other useful natural products—perhaps even among his favorite microbes, the actinomycetes.

Waksman was a rising star in microbiology at a time when researchers were focusing on a ghoulish question. What became of all the microbes that caused deadly diseases—typhoid, dysentery, cholera, diphtheria, pneumonia, bubonic plague, and tuberculosis—when a dead body was buried in the earth? When scientists searched the nearby soil, they found few of these germs, and they concluded that either the microbes could not exist in the soil, because the environment didn't suit them, or they were consumed by predators larger than themselves, or, a far more intriguing possibility, they were destroyed by other microbes.

In London at the turn of the twentieth century, researchers found that the cholera bacterium, *Vibrio cholerae*, survived in clean, deep water but not in surface water containing microbes present in the air. Cholera bacteria disappeared quickly, in a matter of hours, in sewage sludge, and also in seawater. *E. coli* was rapidly crowded out in manure piles teeming with other species of microbes. On these microbe battlefields, researchers in Europe and Australia found actinomycetes to be active warriors, but Waksman was reluctant to become involved. He had no medical training and preferred to concentrate on microbes that were useful in agriculture.

When finances at Rutgers improved in the early 1920s, Waksman became a full-time assistant professor in his chosen pursuit—the microbiology of the soil. The condition of his labs was still pitiful, and he complained to Dr. Lipman. The division of soil chemistry had only two workers but had "three laboratories and three large closets," and his division, soil microbiology, "had four workers and only one laboratory." Recent alterations to the Administration Building had not included painting his walls, which "absolutely demoralizes the assistants and discourages the workers," he wrote to Lipman.

In 1923, Waksman and his graduate assistant, Robert Starkey, saw actinomycetes producing clear zones when matched against other bacteria. "A zone is found free from fungus and bacterial growth," their joint paper concluded, and "numerous" microbes, including bacteria, fungi, and actinomycetes, "bring about injurious or destructive effects upon themselves or upon other soil organisms." But Waksman was interested only in the effects on the fertility of the soil. He did not link this strange activity to the possibility of curing human infectious diseases. "Unfortunately our own observations on the growth inhibiting effect of actinomycetes upon other microbes were not pursued further at that time," he later wrote.

In 1924, Waksman took six months off from Rutgers to go to Europe on a "grand scientific tour" with his wife and their four-year-old son, Byron. It was the first of five European tours that he would make with his wife before the outbreak of World War Two. In 1924, the main attraction was a conference on soil science in Rome organized by Jacob Lipman, who allowed Waksman to continue to be paid his small Rutgers salary but gave him no expenses for the trip. Despite his tight budget, Waksman packed in a hectic schedule of visits to major soil microbiology laboratories in Britain, France, Germany, Sweden, and Holland. In Paris he met and struck up a thirty-year friendship with the Russian pioneer of soil microbiology, Sergei Winogradsky, now an émigré in Paris. And in Holland he visited Martinus Beijerinck, who made his reputation by discovering viruses in 1898 and went on to find bacteria that make nitrogen available to plants. According to Waksman, Beijerinck greeted him with the words "You are the actinomyces man." Waksman also went to Moscow and even his hometown, Novaya Priluka, in Ukraine. There he witnessed the ravages of the revolution and the civil war and saw again the little house where he was born. "It looked like a hole of a troglodyte," he wrote later. He returned to America determined to write source books to fill the gaps in the literature of soil science.

"I was primarily a soil microbiologist," he wrote, "studying soils and composts, peat bogs and manure piles . . . concerned with products of microbes that are used in green plants." He "scarcely dreamed of becoming profoundly involved in problems dealing with human health." He was "too busy completing [his] work on the distribution of different groups of microorganism in the soil, their role in the decomposition of organic matter and formation of humus." His studies resulted in several major works that said almost nothing about the possible medical application of his fighting microbes.

In a 360-page book, *Enzymes*, published in 1926, he devoted only one paragraph to antagonistic bacteria. In his 894-page tome *The Principles of Soil Microbiology*, he wrote only two pages on "antagonism and symbiosis among microorganisms." On another page, he mentioned the "inhibitive effects" of fungi and actinomycetes. In a smaller book, *The Soil and the Microbe*, written with Starkey, now Waksman's deputy, in 1931, they discussed the role of microbes in the life cycle of soil organisms. But he wrote only one sentence about bacteria fighting among themselves.

In his lectures and scientific papers, he would remind his students and readers that the soil was a complex system, our knowledge of it limited, our methods crude, and we were still unable to understand how it works. In the basement lab, his students followed "a semi-military regime," often working weekends during the depression years because they had no money to spend. They wrote brief descriptions of each day's projects in five-by-seven-inch lab notebooks, which Waksman reviewed at the beginning of each week. One student recalled what was known as the "book parade." "Waksman would spot Harry and say, 'Let me see your book.' Waksman would glance at it and add, 'Tell Dave to bring his book.' Harry, disarmed, would go down to the basement lab and pass the word. Dave would submit his book and come back to order another student up to the office. The books were returned when Waksman spotted an error, or something unclear, but he never accused anyone of being a slob and all partings were amiable. The book parade seemed to me a little Teutonic." Waksman rarely visited his students in the basement lab, even then. But once a year he held a spring cleaning, which he obviously enjoyed. All drawers and cabinets had to be open for inspection, and Waksman would walk in followed by his assistant, who carried a laundry basket. The student wrote, "If he found equipment lying on the bench, or chipped, or unlabelled, he would say, 'Vat's this for?' in his Russian accent (which he never lost) and if there was any doubt, he would tell his assistant, 'Throw it in the basket.'"

In those days, Waksman's students were still not looking for medical applications. "The soil and the microbe," Waksman wrote, "await the investigator [who] is not looking for practical gains but for explaining the obscure and observing the unknown. The application will doubtless come." Undoubtedly, Waksman missed a great opportunity. Had he pursued what he had observed with Pasteur's "prepared mind," he, not Alexander Fleming, might have been the first to discover an antibiotic.

But he was not a physician, like Fleming, and Rutgers had no medical department. In his daily life, Waksman was not exposed to faculty discussions about the therapeutic value of "magic bullets" like salvarsan, or the sulfa drugs that followed. In 1932, the German doctor Gerhard Domagk, working at the giant chemical company I. G. Farben, found a bright-red dye that cured mice infected with pathogenic streptococci. This new compound, named Prontosil, was good for fighting a wide range of bacterial infections and later gave rise to the sulfonamides, or sulfa drugs, which had a major impact on the treatment of infectious diseases. (The life of Winston

Churchill was saved by a sulfonamide when he developed pneumonia after the Tehran Conference with Stalin and Roosevelt at the end of November 1943.)

Yet Waksman still lacked funding to expand his research. In America, medical research, like other scientific research, suffered from a lack of public assistance. In the 1920s and '30s, the National Institutes of Health and the National Science Foundation did not exist. Waksman relied on his wits to attract support. He was a good salesman, a scientist-entrepreneur who never seemed short of industry sponsors.

He helped tanners find enzymes for defatting hides; brewers, enzymes to clarify beer. He convinced the local mushroom industry that providing funds to investigate a compost mix of alfalfa, peanut shells, and tobacco stalks was a better bet than relying on horse dung from the declining stables of the Philadelphia police department. These links with industry provided rare funding during the depression years, and endeared Waksman's graduate students to him for providing them with beer and mushroom tastings. There were some unexpected delights. On one famous evening in the college auditorium, female models paraded in then-daring off-the-shoulder evening dresses with fringes of miniature orchids bred in Dr. Waksman's Department of Soil Microbiology. The event was sponsored by a local businessman hoping to sell the orchids at debutante balls.

A few independent foundations gave research grants, and in 1932, the National Tuberculosis Association funded Waksman to study the fate of TB germs in people and animals who died of TB and were buried in the soil. He assigned the task to one of his graduate researchers, who found that the tuberculosis bacteria were greatly reduced in some soils. But Waksman did not follow up this interesting result. Similar results were being obtained by other researchers, and Waksman thought they all seemed to lead nowhere. He was not "yet prepared to take advantage of these findings."

In late 1935, Fred Beaudette, Rutgers's director of Poultry Pathology, brought Waksman a test tube containing a TB bacterium specific to poultry that had been destroyed by a fungus that had accidentally contaminated the tube. This was indeed a "happy accident" of the kind that Fleming had encountered seven years earlier with penicillin. Yet Waksman was still not ready to seize the opportunity, this time staring him in the face.

There was, of course, a perfectly good and understandable reason for not wanting to test a microbe's ability to destroy pathogenic bacteria: the risk of catching the disease. In his writings, Waksman never mentions this as a factor, but it must have been on his mind. His underfunded laboratories were poorly equipped to protect the workers against infections, or even the hazards of handling dangerous chemicals. In the basement laboratory, protective clothing consisted of worn and torn white lab coats and some "very crusty, black, rubber lab aprons designed to catch splashes of hot acid." These coats were "hung on spikes driven into the wall" when not in use.

BUT WAKSMAN COULD not ignore the research coming out of Europe and Russia. In the mid-1930s, the Russians led the world on research into the antagonistic properties of Waksman's precious actinomycetes. By 1935, Russians had published four papers on the subject; Waksman had published none. A key Russian paper reported that actinomycetes were antagonistic to *Bacillus mycoides*, one of the standard bacteria tests for antibiotics. The paper concluded, "The question of interrelationships of soil microbes deserves profound research." Waksman had an enormous advantage over his peers in America in being able to read these papers, not just the English summary that was always included but the whole paper, and some have speculated that the Russian research started to turn his mind toward the possibility of antibiotics, a suggestion he never acknowledged.

In 1936, at the Second International Congress for Microbiology in London, Alexander Fleming discussed the antibacterial properties of his penicillin, a debate which Waksman later listed as an important event in the evolution of his own thinking. One of Waksman's graduate students recalled that Fleming's discussion was when Waksman became "seriously interested" in antibiotic research.

Later in 1936, Waksman began to study the published papers on warrior microbes and wrote two papers for *Soil Science*, the journal started at Rutgers by Dr. Lipman. The first paper reviewed the current literature, including the four Russian papers. A measure of Waksman's absence from basic research in this area is that of the 107 papers he listed, only 2 were written by him.

In the second paper, also finished in 1936, Waksman and a graduate student, Jackson Foster, tested a fungus, a bacteria, and an actinomycete from a Scottish peat bog. They were all capable of producing "substances which are antagonistic" to other soil microbes when grown in petri dishes containing artificial nutrients. In 1937, another Russian researcher found that antagonistic actinomycetes were "widely distributed" in different soils in the Soviet Union. Of eighty cultures isolated from various soils, forty-seven possessed antagonistic properties, but only twenty-seven were found to be capable of liberating toxic substances into the nutrient agar on a petri dish.

In 1938, Waksman was especially influenced by the work of one of his former students, René Dubos. A Frenchman who had qualified in agriculture and immigrated to America in 1924 after hearing Lipman speak at the conference on soil science in Rome, Dubos worked for his Ph.D. under Waksman at Rutgers. He discovered a soil microbe that produced an enzyme capable of breaking down cellulose, the key ingredient of plant stalks and tree bark, and turning it into plant food. Similar work was being carried out at the Rockefeller Institute for Medical Research, in New York City, where Dubos later moved. There, he eventually isolated a bacterial enzyme that destroyed the sugary coat of the bacteria that causes pneumonia. Unfortunately, the enzyme was too toxic to be used by humans suffering from pneumonia, but Dubos was sufficiently encouraged to begin the first systematic search for antibiotics in the soil.

In 1939, he found an antibacterial agent produced by a bacterium, *Bacillus brevis*, and named it tyrothricin. The Rockefeller biochemist Rollin Hotchkiss helped him recognize that it was made up of two compounds, tyrocidin and gramicidin. Tyrocidin was toxic to mice, but gramicidin cured experimental infections in mice, without side effects. Gramicidin was too toxic to be administered to humans intravenously, but it was effective when used on open wounds. The Russians produced their own version of Dubos's discovery, known as gramicidin S (for Soviet), and used it throughout World War Two as their main antibiotic.

That same year, the Russians struck again. Two researchers, N. A. Krassilnikov and A. I. Korenyako, again found that many species of actinomycetes produced antibiotics. The Russians concluded that "one cannot escape the possibility of using the bacterial factor of actinomycetes" to treat bacterial diseases. For the first time, they discovered two that were active, ever so slightly, against *Mycobacteria*, the group that causes tuberculosis. To anyone searching for a cure for TB, it was a powerful clue that such an antibiotic might be found.

By late 1939—in the wake of pioneering research by the Russians, a major discovery by Dubos in New York, and the beginning of the war in Europe—Waksman, the soil microbiologist who had pledged his life to microbes that could be used in plants and industry, was finally ready to change the direction of his research to look for antibiotics to cure human diseases. All he needed was a sponsor.

4 · The Sponsor

TWELVE MILES DOWN THE RAILROAD TRACK from the Rutgers campus is Rahway, New Jersey, once an old Indian settlement and a stop on the stagecoach run from New York to Philadelphia. Since 1903, Rahway has been the home of Merck & Co., then and now one of the most important pharmaceutical concerns in the country. Friedrich Jacob Merck opened the original family-owned apothecary, the Engel-Apotheke (Angel Pharmacy), in 1656 in Darmstadt, Germany. In 1827, the Merck company started producing morphine, codeine, and cocaine. By the 1890s, Merck was selling so many products in America that the family dispatched the twenty-four-year-old George Merck to set up shop there. He settled in Manhattan, bought 150 acres of Rahway, and later sent his son, George Jr., a blond, blue-eyed giant at six feet five inches, to Harvard. On George Sr.'s death in 1926, his son took over the business.

By the beginning of World War Two, Merck was also producing vitamins. First came vitamin B1. Until the Merck chemists figured out how to synthesize the compound, tons of rice bran went into one end of Merck's Rahway plant, and fractions of an ounce of vitamin B1 came out the other end. Soon there was vitamin B2 for pellagra, vitamin B12 for anemia, vitamin C for colds, and vitamin A for eyesight. George Merck was also keeping a watchful eye on the Rutgers Department of Soil Microbiology. Like many other microbe researchers, Selman Waksman was experimenting with ways to use fungal fermentations to make citric acid, used in foods and soft drinks, and fumaric acid, used in dry cleaning.

At the beginning of 1939, Merck engaged Waksman as a consultant on

ed the Nazis in

other proposal.
of another $150
s." "I informed
r. "They placed
Merck in this
riological and
tion and eval-
s of tests that
ilities. In ex-
ew drug that
a royalty of

omising re-
ger to start
er constant
enicillin to
ompanies
erle Labo-
d the Up-

hre
ack of sta
production a
Penicillin wa

and then $150. In the summer
student fellowship specifically
program was successful, and
renewed on an annual basis.
at his consultancies with Merck,
the funds went into fellowships
some for "collaboration" between
private consulting" between him-

essed interest in hiring Waksman
h in those days the word antibiotic
led them "antibacterial chemothera-
enicillin. Although Alexander Flem-
cterial powers in London in 1928, he
concentrating, or purifying it, and it
.

out Fleming's discovery, and in 1933 a
State College had studied Fleming's mi-
confirmed Fleming's claims about the
unable to extract it himself, made no fur-
American drug companies, Eli Lilly and
s potential. Squibb researchers carried out
produced a well-reasoned statement, now
cillin, concluding that "in view of the slow
and slowness of bacterial action shown by
marketing as a bactericide does not appear
delined in favor of the readily available sulfa

6, a chemist at M ck was shown a culture of Fleming's *Penicillium* n by a physician from New York's Beth Israel Hospital who predicted ore interesting antibiotics were on the way. Three years later, Merck's arch director, Randolph Major, asked Dr. Waksman's advice, and he gested taking penicillin seriously. Other similar agents would probably on be found, he forecast. Merck immediately hired three new staff members to "study isolation of therapeutic substances from micro-organisms."

In Britain, the start of the world war had revived interest in Fleming's penicillin. At Oxford University, Howard Florey, an Australian pathologist,

and Ernst Chain, a German Jewish chemist who had fl
1933, began work on purifying penicillin.

In the fall of 1939, Merck returned to Waksman with ar
This time the company offered him a second consultancy—
a month—for information about "chemotherapeutic agent
them of my own interest in antibiotics," Waksman noted late
another fellowship in my laboratory and engaged me to help
field of research." Merck agreed to carry out "chemical, bacte
biological tests for the production, purification, plus identifica
uation and to arrange for clinical trials." These were the kind
could not be carried out at Rutgers because of a lack of fac
change, Merck would have the exclusive right to develop any n
resulted from the research. The company would pay Rutgers
2.5 percent of net sales.

In August 1940, the Oxford team published their first pr
sults of the use of penicillin on ten patients, and the team was e
development. But British industry was overstretched, and und
air attack. Florey and a colleague, Norman Heatley, brought p
America and found a government not yet at war, and drug
eager to be the first in the antibiotic market. Merck, Squibb, Le
ratories, and Pfizer & Co., in the East; Abbott, Parke, Davis, ar

George Merck of Merck & Co. with vial. (The Merck Archives, 2011)

john Company in the Midwest. Merck agreed to be part of a massive, U.S. government–sponsored war effort to produce penicillin. George Merck sent a telegram to Vannevar Bush, director of the Office of Scientific Research and Development: "Command me and my associates . . . if you think we can help you." The Roosevelt administration launched an astonishingly successful example of government-science-industry cooperation, second only in wartime to the atomic bomb project. It would eventually involve ten American and five British firms, combining efforts to make the drug for Allied troops.

WAKSMAN'S DEAL WITH Merck caused quite a stir in the offices of the Rutgers administration. They wanted to make sure the university got its share. Like many universities of the day, Rutgers had no policy for dealing with faculty who made patentable discoveries. The most recent case, in 1933, concerned a professor of pomology named M. A. Blake, a well-known breeder of peaches who was called the father of the New Jersey peach industry. He was especially proud of his latest nectarine crosses and wanted to apply for a patent.

Whether Blake himself had the right to take out a patent depended on his contract, the Rutgers lawyers advised. If he had been employed specifically to breed nectarines, he would have to assign the patent to Rutgers, but if he was a "general employee" in the fruit-breeding department, and he had bred this spectacular new nectarine in the course of other work, then he would be entitled as an individual to apply for a patent and collect royalties. The lawyers noted, however, in view of Rutgers's duty to make agricultural discoveries available for free to the public, that Professor Blake "might be embarrassed" if he started to profit from the patent. In that case, there was a third way: He could assign it to the nonprofit Rutgers Endowment Foundation, a body originally set up to receive alumni donations. A percentage of the royalties could be paid to the professor, the lawyers said, in line with similar arrangements at other universities.

This quickly became Rutgers's policy; the only question was what percentage, if any, of the royalties to allow the discoverer. In 1937, Rutgers agreed to a 50-50 split—until it found out that it was being overgenerous compared with other universities, or, as the Rutgers comptroller,

A. S. Johnson, observed, that it had been "decidedly off on the wrong foot." Rutgers reduced the discoverer's share to 25 percent, but even that was above what other institutions were paying; Purdue's was fixed at 20 percent, the University of Wisconsin's at 15 percent. Wisconsin's Alumni Research Foundation director, A. L. Russell, advised Johnson to "keep in mind" that university patents are "to be taken out primarily in the interest of the public rather than for the inventor."

While the Merck deal with Waksman was being worked out, the first of Waksman's graduate students to work on the antibiotic project arrived at the Department of Soil Microbiology. Boyd Woodruff, a tall, confident, genial farmer's son from South Jersey, joined Waksman's laboratory in July 1939. His parents were determined that he should have a university education, but all they could afford was the state-supported agricultural and engineering course at Rutgers. He lived with other students above the chicken house, which at that time accommodated 125 white Leghorns. Woodruff earned pocket money selling farm eggs.

Woodruff had found the college experience exciting and sometimes a little overwhelming. He had gone to a concert for the first time, and had celebrated with his fellow students all night in 1937 when the Rutgers football team scored its first victory, 29 to 27, over Princeton since 1889. He graduated in soil chemistry, and Waksman offered him a college fellowship of $900 a year—20 percent more than fellowships elsewhere. The money came from Merck's generous contributions to Waksman, now totaling $3,600 a year.

As a result of his European tours, Waksman attracted students worldwide. Eleven graduate students crammed into the two upstairs laboratories of the Department of Soil Microbiology. They came from China, South America, Europe, and across the United States. At first Woodruff was "terribly discouraged" when Waksman put him to work on composts and gave him little direction. He relied heavily on Waksman's deputy, Robert Starkey, as had most of Waksman's graduate students over the years. As one of them recalled, Starkey was their "great provider of materials and receiver of complaints—the equivalent of an assistant in a steel mill. He remembered; he got things done. He told us how to make our cases to Dr. Waksman. Modest to a fault, totally loyal to the Department—it's inconceivable that Dr. Waksman could have operated without him." Woodruff worked alongside a visiting student from China who was trying to find out the minimum temperature needed to kill all the harmful bacteria in human feces so that it could be used for compost. Since human feces were not used for compost in America, Waksman suggested that Woodruff should study horse dung, horse urine, and straw, and see how the combination worked. After that, Woodruff moved on to potato scab disease, a serious problem in New Jersey, caused by an actinomycete. The microbe does not grow under acid conditions and is controlled by farmers by adding sulfur to the soils. A bacterium oxidizes the sulfur to sulfuric acid, thus producing the desired soil acidity.

One day, toward the end of 1939, Waksman received word from Merck that the Oxford team was successful in isolating and purifying penicillin. Highly agitated, Waksman appeared in the laboratory and told Woodruff to "drop everything. See what these Englishmen have discovered a mold can do. I know the actinomycetes can do better." Certainly, the Russians had already suggested that the actinomycetes were worth testing.

Waksman took Woodruff to see René Dubos at the Rockefeller Institute so that he could learn the so-called soil-enrichment method that Dubos had used when he discovered tyrothricin. In this method, the researcher adds a disease-producing bacterium to pots of soil over two to three months, hoping this will favor growth of the species of actinomycetes that can kill and feed on that particular bacterium. Woodruff chose *E. coli* as his disease bacterium, adding billions of cells to pots of Rutgers college farm soil he knew was rich in actinomycetes. Each week he counted the surviving *E. coli* by taking a sample of the soil in the pots and growing it on nutrient jellied agar in a petri dish. It was easy to spot the *E. coli*, which appeared as clumps of mold with a distinctive metallic luster. Each time he counted them, the number was reduced until at the end of the three months Woodruff's pots of enriched soil had, as he put it, "become highly efficient *E. coli* killing machines." He isolated the sturdy-looking cultures of *Actinomyces antibioticus*, which produced a red chemical substance apparently responsible for the killings. He had found a new antibiotic. Woodruff and Waksman named it actinomycin.

Antibiotics are often difficult to extract and purify, as Fleming had found with penicillin, but actinomycin was relatively easy. Woodruff and Waksman hoped they had struck lucky, but now they needed to make enough actinomycin to study its effect on disease in small animals.

Waksman took a sample to Merck, where the new drug could be produced on a larger scale than in his laboratory. Then he and Woodruff wrote up their results in a paper that was published in the spring of 1941. Merck's researchers purified and tested actinomycin in animals. The results were shocking. It certainly killed disease bacteria, but it was so toxic it also killed laboratory mice in twenty-four to forty-eight hours. There was no question of testing it on humans. Rat poison was about all it was good for.

Waksman was "truly excited" about the discovery, however. "Once I had an actinomycete in my hand which produced antibiotic activity, everything changed," Woodruff recalled. "Waksman started coming into the lab immediately after lunch each day."

At this early stage, these microbial chemical weapons were so new they had no specific name. They were known as "antibiotic substances." The term came from "antibiosis," meaning "against life," a word first used by Louis Pasteur's pupil Jean Paul Vuillemin in 1889 to describe the antagonistic effects of microorganisms. By the 1930s, the use of "antibiotic" as an adjective, as in "antibiotic substance," was fairly common in the scientific literature, especially among European researchers. Waksman himself used it. But now a proper noun was needed.

At one of Waksman's Friday afternoon meetings when students discussed the latest scientific journal articles, Waksman asked the gathering for a noun to describe the new wonder drugs. The students offered some suggestions, but Waksman already had an answer: "antibiotic." He even had a definition ready. An antibiotic, he said, was "a chemical substance produced by microorganisms, which has the capacity to inhibit the growth of and even destroy bacteria and other organisms." The noun entered the medical lexicon, but Waksman's colleagues would argue furiously whether he should be credited with "coining" the word, or simply applying it as a noun. The debate continues.

Now armed with the redefined word, Waksman put his students to work, screening soils and composts for the new candidate antibiotics. One of these researchers, a postdoctoral student named Walter Kocholaty, arrived at Rutgers from the University of Pennsylvania to learn the methods Waksman was using. Instead of soil enrichment, which took up to three months to produce a likely microbe, Waksman now switched to a second procedure to save time. Tiny amounts of farmyard soil, or compost, were diluted in tap water and drops of the soil and water mix put into a petri dish containing living cells of a disease bacterium in jellied agar. The disease bacterium was the only source of food for the microbes from the soil and water mix. As these microbes grew, the ones capable of killing and eating the disease bacterium created the tell-tale clear zones in the agar. Kocholaty found a strain of a purple-colored actinomycete, A. *lavendulae*, produced good clear zones in the petri dish.

Waksman had first come across A. *lavendulae* in his early research in 1915 and Kocholaty now isolated robust-looking strains from the colonies growing in the petri dish, transferred them to separate dishes, let them grow, and then tested the strain against more disease-producing bacteria by the "cross-streak" method. This involved "streaking" a line of the A. *lavendulae* strain down the right-hand side of a petri dish and then streaking the bacteria to be tested at right angles to it—like bringing troops up the front line. When clear zones appeared in the lines of disease bacteria, Kocholaty knew he had a new antibiotic in A. *lavendulae*.

His time at the lab was over, however, and Waksman gave the strain to Woodruff for further tests. He grew the strain of A. *lavendulae* in various different broths to see which one would cause the strain to produce the most antibiotic. Waksman and Woodruff named the antibiotic streptothricin, after *Streptothrix*, or "twisted fungus," the name given to an actinomycete by a nineteenth-century German researcher. Woodruff made enough streptothricin for animal tests in the only lab available at the Rutgers college farm—the dairy. There, they found it seemed to cure cows of a bacterial disease, brucellosis, that caused contagious abortion. Excited, Waksman and Woodruff now took their second antibiotic to Merck, where company researchers confirmed their findings. Streptothricin looked like a winner. For the last six months of Woodruff's Ph.D., Waksman sent him to work on streptothricin at Merck, where he could be Waksman's "eyes and ears" as the company continued its testing.

Merck was so excited by streptothricin, the company even built a small production plant. It cured bacterial infections in mice, but when company researchers started a clinical trial on four human patients, they found that almost immediately the patients' kidneys stopped producing urine, so the treatment was stopped. "We thought they were all going to die," Woodruff recalled. "We were all worried for a couple of days." Woodruff was devastated. The failure of streptothricin was the greatest disappointment of his life, he would say later. But he was fortunate; Merck offered him a job and he stayed at the company until retirement.

Despite the two failures of actinomycin and streptothricin, Waksman was determined to keep his search going. The ranks of his students available to do the screening were dwindling, however. The draft board was now taking staff from him, mostly men. When Woodruff arrived at the department there were eleven graduate students; by the middle of 1942, there were only four. One of them was a skinny youth, a recent Rutgers graduate named Albert Schatz.

By the time Schatz arrived in the early summer, the graduate students had isolated some 400 cultures of actinomycetes and 160 of fungi and were concentrating on four microbes that produced antibiotics: two from actinomycetes, actinomycin and streptothricin, and two from fungi, clavacin and fumigacin.

Schatz had wanted to do his Ph.D. on soil science but the department head was not as good a salesman as Waksman, didn't have a commercial product, and could not attract any funding. Schatz lived for free in a room off the Plant Pathology greenhouse in return for minor duties tending the experimental plants grown to fill President Roosevelt's Victory Gardens. Like Boyd Woodruff, he lived off eggs from the poultry shed and milk from the dairy.

Waksman's reduced team of four graduates was happy to work on antibiotics under its "fatherly professor," a short man, by now slightly overweight, with a bristly mustache and bright, intelligent eyes. His clothes were never pressed, and his vest always carried some evidence of a recent meal. The young students, like Schatz, admired Dr. Waksman, even idolized him, referring to him—only behind his back, of course—as "Old Waksie," "the Boss," and sometimes "the Great White Father." "The atmosphere was terribly stimulating," recalled Doris Jones. A male former student recalled, "One of his [Waksman's] favorite tricks was to stop you in the hall and ask how everything was coming along. As you started to explain something, he would say, 'Good, let us not waste any time, follow me.' He would then take you into the men's room where you continued your scientific report while he was not wasting any time." His European colleagues regarded him as "a typical representative of the Russian Jewish intellectual at its best." Schatz started work immediately on experiments on streptothricin and another antibiotic, clavacin, that had been found to be toxic in animal tests.

The war was taking its toll on Waksman's team. By July, all except Schatz had left to go into industry or teaching, which offered a draft deferment—Boyd Woodruff had gone to work at Merck, for example—or had been drafted into special branches of the military where their lab experience could be useful. Waksman's antibiotic project was struggling for lack of funds. Seeing that the federal government was funding the production of Fleming's penicillin, and now Dubos's gramicidin, he applied to the government's Committee on Medical Research, which had been set up to fund wartime medical projects. But he was turned down. The committee was looking for projects that could produce results "within relatively short periods of time." Penicillin and gramicidin had already proved their worth. Waksman's project covered "a broader field" that was, at that moment, more theoretical than practical. The committee suggested that Waksman try philanthropic foundations, and it passed his application to the Commonwealth Fund, in New York. He was soon awarded a grant of ninety-six hundred dollars—more than enough to add equipment and research staff, if he could find researchers exempted from war service.

In October 1942, after five months of working for Dr. Waksman, Schatz received his draft papers. Waksman tried unsuccessfully to persuade the draft board that Schatz was "more valuable" working on the search for new antibiotics than "becoming a private in the army." He then pleaded with the draft board to at least give Schatz a job related to his studies, and in November 1942, Schatz became a laboratory technician at an Army Air Corps hospital in Miami Beach. There, thousands of miles from the front lines, he was exposed to the terrifying effects of war. He drew blood from seriously wounded and diseased soldiers brought home to America, and he identified the microbes responsible for their infections.

It was relatively simple work, easily accomplished, as he was already something of an expert. But Schatz quickly learned that once he had identified the germs, the army often had no medicines to cure the diseases. Penicillin was being produced as a war priority but was still in short supply, and its healing powers were strictly limited. It had no effect on the germs that cause illnesses epidemic in wartime, such as cholera, typhoid, urinary and intestinal infections, and, the persistent killer, tuberculosis. Schatz knew which soldiers in the military hospital had no hope of recovery, and in his off-duty hours he volunteered to sit by their bedside, comforting them as best he could in their last hours.

As news from the front improved in the spring and early summer of 1943 and the Allies were preparing to invade Italy, Schatz was hopeful of an early return to his studies. He was homesick for the camaraderie of Dr. Waksman's close-knit group. He wrote to Waksman as frequently as his duties permitted and complained about the monotony of military life, saying that he hoped to return to his studies soon.

"All of us are eager for the war to end," he wrote in one such letter. "I personally can hardly wait to get back home and return to school. Regards to all, Very respectfully yours, Pvt. Albert Schatz. P.S. I hope all of Africa is in Allied hands by the time this letter arrives." Waksman, always encouraging, suggested he should look for antibiotics in the Florida soils, and Schatz mailed samples of promising candidates to Waksman for his microbe collection.

His return to Rutgers came even sooner than he had hoped. As he was loading supplies onto a truck, he twisted his back, and army doctors discovered he had more than a sprain. Private Schatz had a malformation of the spinal cord, apparently undetected during his draft board medical exam. He was declared unfit for military service and honorably discharged. The army recommended him for a Good Conduct Medal, which he never

Pvt. Albert Schatz, U.S. Army Air Corps, Medical Detachment, in Miami Beach, Florida, in 1943. (Courtesy Vivian Schatz)

actually received, and gave him the train fare home to New Jersey on June 15, 1943.

Waksman quickly reenrolled him in the antibiotics program, but he could only pay forty dollars a month, a third of what Schatz had received before he went into the army. Schatz didn't mind. Despite his back injury he was, if anything, more self-assured, more independent, and he returned to his free room off the Plant Pathology greenhouse. He was ready to start working round the clock, sleeping "in his cloth," as he put it.

5 · A Distinguished Visitor

IN SCHATZ'S EIGHT-MONTH ABSENCE, WAKSMAN'S TEAM, which now included two new women graduates—Doris Jones and Betty Bugie—had not found a single new antibiotic. But they had had lots of fun trying, according to Jones. She loved "the close-knit group, the Friday brown-bag lunches where the smell of sandwiches mixed with the scents of the molds, especially the earthy odor of the actinomycetes." A bright, jolly, optimistic person with a self-deprecating wit—she once described herself as having "well-padded bones" and acknowledged her nickname, "Moose," because of her loud voice—she didn't mind the hard work. "We call the laboratories the Salt Mines because in order to pull a practical antibiotic producer out of Mother Nature we literally have to 'work our asses off,'" Jones recalled later. But she appreciated how Dr. Waksman urged his graduate students not "to waste time studying extraneous things in books. I worship him."

Merck researchers were testing the most promising product so far—streptothricin—but it was becoming apparent to Waksman that streptothricin's toxicity could not be sufficiently reduced, even in the purer form made by the Merck chemists, to "offer hope for its eventual usefulness." So when Schatz resumed working on his Ph.D., the opportunity to make his mark on medical history was wide open. And Waksman, for the first time, was interested in testing his microbes against the genus *Mycobacterium*, some species of which can cause tuberculosis.

A year earlier, he had been pressed to start such a project by his son,

Byron, who was then finishing his degree in bacteriology at the University of Pennsylvania Medical School. In a letter, Byron had offered to do a summer project at Rutgers, first testing microbes against the nonvirulent strains of *Mycobacteria* in Waksman's lab and then, if an active agent was found, testing it against the virulent human form in animals. As Waksman did not have animal-testing facilities in his department, that risky second test would have to be done elsewhere. But Waksman was still not ready. "The time has not come yet," he replied to his son.

In their competing versions of what happened next, both Waksman and Schatz later claimed that they were the one to first take the TB project seriously. Waksman said he finally decided to go ahead with the project "early in 1943." Schatz said that on his return from the army in June he knew exactly what he wanted to do for his Ph.D., and he chose the most ambitious research project that any Waksman graduate had ever suggested. He would find a new antibiotic that would cure the diseases his military comrades had died from, and that included tuberculosis.

The evidence does not resolve this disagreement, but seems to favor Waksman. He was certainly moving closer to testing his antibiotics against the TB germ. On June 1, according to his expense records, he went to New York City to meet Dr. Leroy Gardner of the Trudeau Sanatorium at Saranac Lake. The topic was "the problem of bacteriostatic substances in relation to tuberculosis." On June 18, Waksman wrote to Dr. Florence Seibert at the Henry Phipps Institute in Philadelphia, where they carried out research on TB. Dr. Seibert had invented the first reliable tuberculosis test for humans. Waksman asked for fifty to one hundred grams of dried cells of the human TB H37 germ, and also a culture which could be used for growing the organism. Dr. Seibert sent one of each, but warned that the culture was old and of uncertain viability—the dried cells had been alive nine years before being dried in 1943. In other words, she did not know whether these strains would produce meaningful results if used as a test for TB.

Schatz was discharged from the army on June 15 and officially started work in Waksman's department on June 30. Whether Waksman's letter to Dr. Seibert was prompted by a conversation with Schatz after his discharge from the army is not known.

In any case, Schatz did not begin his experiments testing his candidate antibiotics against *Mycobacterium*; that was to be the second part of his Ph.D. He started by enriching soil in pots with *E. coli*, as Boyd Woodruff had done to find actinomycin.

Schatz's first experiment, as noted in his 1943 lab notebook, was on June 30 and dealt with a "general survey of the occurrence of antagonistic microorganisms." He isolated bacteria, fungi, and actinomycetes which might be responsible for destroying *E. coli*, then tested them by the streak test to see whether they produced the clear zones. On July 23, he noted in his lab notebook, he gave up the soil-enrichment method and instead switched to the agar plate method used by Dr. Kocholaty when he found streptothricin. This meant random testing of soil samples against known disease-causing bacteria. In experiments started at the beginning of August, Schatz noted that "some molds are apparently antagonistic immediately upon isolation, but they seem to lose this property upon repeated culture on artificial media." In other words, he still hadn't found anything worthy of isolation and further experiment.

Every researcher knew that no matter how great the effort, luck was always involved in discovery, in biology especially. Alexander Fleming discovered penicillin in 1928 when a fungus spore fell by a happy accident into one of his petri dishes and he noticed the clear zones of antagonism. Too many researchers went on testing for years and never found a microbe capable of killing off a disease. Only the single-minded, obsessive researcher, the kind willing to give up a full life, the "true devotee to science," in Einstein's terms, would have a chance of success. Albert Schatz was such a researcher. Fleming used to describe his microbe experiments as "playing about," and Schatz knew the feeling well. He loved spotting a likely microbe, one that he thought might make a good candidate to produce an antibiotic. He loved fussing over his precious molds as they blossomed into striking, beautiful sculptures of red, blue, yellow, and gray-green. He was so fascinated by the potential power of his friendly bacteria that it seldom felt like work; even the routine, the drudgery, seemed like play. No great skills were needed, he was the first to admit. At this early stage, it was just about a steady hand and a good eye. The basic techniques, which he had learned quickly, were known as "silly simple" by one of Dr. Waksman's graduates.

Schatz already understood that his chances of finding a useful antibiotic were remote at best, and the human TB strain posed a special prob-

Waksman and Schatz working in the laboratory at Rutgers. (Special Collections and University Archives, Rutgers University Libraries)

lem. There was a good reason why others had failed. Of the disease-causing microbes, the deceptively simple cell of *Mycobacterium tuberculosis* presented one of the greatest challenges.

Bacteria are divided into two groups on the basis of their reaction to a stain first used in 1884 by a Danish bacteriologist, Hans Christian Gram. Those cells that retain the stain are called Gram-positive; those that don't, Gram-negative. The distinction is important for diagnostic purposes and is due to a basic difference in the cell structure. Gram-negative cells, which cause typhoid and cholera, have an extra outer layer, making them tougher for antibiotics to penetrate. Penicillin, for example, is effective against Gram-positive bacteria such as *Staphylococcus*, the common cause of blood poisoning, but not against Gram-negative *Salmonella*, which can cause typhoid, or *Vibrio cholerae*, which causes cholera. The TB microbe is even more difficult to penetrate, having an extra layer of protection, a waxy wall preventing the entry of unwanted chemicals.

Schatz was determined to find antibiotics which destroyed the Gram-negative bacteria and that also destroyed the TB germ. After ten experiments he had no promising results, so that was when he switched to a third method—random selection. This is the least complicated of the three methods, relying to a greater extent than the other methods on chance. He took samples of soil, compost, or stable manure, added tap water to make a liquid mix, then put drops of the mix on a petri dish of agar containing a food microbes are known to thrive on—egg albumin. Then he incubated the dish and watched the microbes grow. By mid-October, Schatz had selected two strains from the gray-green *Actinomyces griseus*. One strain, 18-16, excreted an antibiotic active against the Gram-negative bacteria *E. coli*. Another strain came from the culture of the chicken's throat given him by Jones. It was also active against *E. coli*. Schatz named it D-1, for Doris. On October 19, Schatz sealed the test tube containing the 18-16 strain and gave it to his mother.

On October 20, Schatz began isolating his new antibiotic in Experiment 25—the "Collection of Active Material of 18-16." The entry ends with a note: "Material taken over by Dr. W. & E.B." E.B. was Elizabeth Bugie, who was working under Dr. Waksman in the upstairs lab. She began testing the strain in different mixes of nutrients to find the best one for producing the new antibiotic. The favorite seemed to be a meat extract.

It was "impossible to set down in words the excitement that prevailed in the laboratory during those ensuing days," according to the book *Miracles from Microbes: The Road to Streptomycin* by Samuel Epstein and Beryl Williams, who had already collaborated on several books for young readers. The book was published by Rutgers University Press in 1946. Schatz did not agree with the Epstein and Williams account—at least of his mood, which was not excitable, he said, because of the work yet to be done. Many years after he wrote in the margin of his copy of the book, "Not true, the results were *in vitro* tests. Nobody knew the toxicity." He was right, of course. No one could know, until tests on animals were done, whether his new antibiotic would be toxic to humans like others produced so far in Waksman's lab.

Waksman gave a sample of the new antibiotic to Doris Jones in Poultry Pathology for the first animal tests—on chick embryos, still in the egg, infected with fowl typhoid that would normally kill them. Jones injected them with five to ten milligrams of the new antibiotic and many of them hatched. She was so new at this technique that she had never watched them peck their way out of the shells. Then she had to kill them for an autopsy to see if the typhoid bacteria had been completely destroyed. Before the operation, she hosed down the walls of the autopsy room, she recalled later. It was the only method she had of trying to rid the autopsy room of dust-containing bacteria that might have contaminated her experiment. "But I'm paralyzed, I can't squeeze the scissors through the necks of those little beings," she wrote. "Dr. Beaudette mocks me. He won't help. It takes days. Finally, I squeeze and the autopsy proceeds as tears run down." The new antibiotic had worked; no bacteria were left in the chick's organs or blood.

By this point Waksman's lab must have been buzzing with word of the discovery, and the question was what to call the new antibiotic. The naming of it would be the cause of yet another unresolved disagreement between Waksman and Schatz. Each claimed that they had thought of the name first. Waksman was the first to use the name streptomycin in a document, according to the archives. In a letter, he informed Randolph Major, Merck's research director, of the discovery on October 28, saying it could "tentatively be designated as streptomycin." Schatz claims that he was always going to call his discovery streptomycin, but the name does not appear in Schatz's lab notebook until December 14.

Toward the end of October Schatz started to test his new antibiotic from the two strains, 18-16 and D-1, against the harmless strains of mycobacteria in the department's collection. On November 8, he began Experiment 29, which he titled "Bacterial Action of 18-16 Concentration Upon TB." A week later, on November 18, he started Experiment 30, which was designed to test D-1 on TB. The results were promising, but the tests were again against TB from the department's collection and, therefore, nonpathogenic. The real test was yet to come.

AS SCHATZ BEGAN his Experiment 30, a distinguished visitor arrived at the Department of Soil Microbiology. He was Dr. William Feldman, a veterinary pathologist at the Mayo Clinic in Rochester, Minnesota, one of the most famous private clinics in America. Its founder, William Worrall Mayo, a British immigrant, had opened the clinic as a frontier practice in 1846. By the 1930s, the Mayo Clinic was well known for its work in trying to find a cure for tuberculosis. Feldman and a colleague, Corwin Hinshaw, were the principal researchers.

Feldman had immigrated to the United States from Scotland in 1894 at the age of two, and he had grown up in Colorado knowing about TB. His mother had recalled, from her own childhood in Glasgow, the suffering caused by the disease. For his doctorate at Colorado State University, Feldman had taken part in a nationwide effort to eradicate TB in cattle, a cause of often-fatal tuberculosis in children. He had also written a book on tuberculosis in birds. His partner, Hinshaw, was also a Westerner, a zoologist and medical doctor with expertise in parasites and bacteria. He had been at the Mayo Clinic since 1933, working on pulmonary diseases, especially pneumonia. Their mix of disciplines was excellent for testing new drugs on animals; they made a perfect team.

For more than a decade they had tried a variety of drugs, including compounds of gold and arsenic and the latest sulfa compounds, on guinea pigs infected with the human TB strain. Results from the sulfa drugs were encouraging. They treated about a hundred patients with drugs that showed unmistakable promise. One of the sulfas, Promin, gave the first hint that the waxy outer wall of *M. tuberculosis* could be breached, but did not destroy the germ. Nothing seemed to halt the onward march of the TB microbe, which nestled in inaccessible places in the lungs of patients suffering from the disease. Feldman felt that he and Hinshaw had a "foot in the door," as Feldman put it, but their peers thought they were "wasting their time"; the sulfa drugs might halt the infection, but would never completely destroy TB. This only made the two scientists more determined and stubborn.

They devised more rigorous tests, insisting that virulent and nonvirulent TB germs had different characteristics and that any new drug had to work against the virulent type before they would even consider animal tests. They used guinea pigs because they are highly susceptible to human TB and, once infected, are severely hit by the disease; if a drug worked, it should also work in humans.

When Feldman heard about Waksman's new research into antibiotics, he asked for samples, first of the fungus-produced clavacin, which sounded promising. Waksman was cautious. All four of the antibiotics discovered in his lab—actinomycin and streptothricin from the actinomycetes and fumigacin and clavacin from fungi—were too toxic. Even so, he invited Feldman to visit Rutgers.

Feldman spent several hours talking with Waksman in the upstairs laboratory, offering to carry out "co-operative studies" if and when Waksman's lab produced a new antibiotic. Waksman said he would let Feldman know. Before they parted, Waksman introduced Feldman to Schatz, but only briefly, and he never mentioned Schatz's new discovery of streptomycin. Waksman's loyalty was to his sponsor, Merck.

ON DECEMBER 14, Schatz began Experiment 33, using extracted streptomycin on an H37 strain of TB, apparently from the strains of uncertain viability Dr. Seibert had sent Waksman back in June. The results were again promising.

Schatz was astonished to have found not one but two likely candidates so quickly. What the future held for his discovery, however, no one could really say. His chosen two might be eliminated as quickly as the other four because of toxicity. And while he assumed that Dr. Waksman would do all he could to make his discovery a reality and give him full credit for it, sometimes Schatz was not quite so sure.

Waksman's permanently wrinkled attire gave the impression of an eccentric academic absorbed in lofty scientific principles and novel ideas, a professor dedicated to pure research. And his passion for the little-known actinomycetes added to the image of unworldly benevolence. However, Schatz and the other researchers knew another side of their professor.

Selman Waksman was the best-organized, the most practical, and the best-connected professor at the College of Agriculture. Beneath his bonhomie, the twinkle in his bright eyes, the Friday brown-bag lunches, and the informal summer picnics on the Jersey Shore with "kosher" hot dogs wrapped in bacon, there was a traditional European department head who followed a rigid code of rank when it came to his apprentices.

One unsettling story about Waksman's relationship with his students had become a legend at Rutgers. Oddly, it involved another Russian Jewish immigrant, Jacob Joffe, a fiery character who was born in Lithuania and had immigrated to America on the eve of World War One. He became one of Waksman's first graduate students. He finished his dissertation on

Waksman cooks "kosher" hot dogs wrapped in bacon at a Department of Soil Microbiology picnic in 1945. (Courtesy Vivian Schatz)

a bacterium that had the remarkable ability to turn sulfur compounds into sulfuric acid, which could then be used to free up phosphates in the soil, a natural fertilizer. This was a major breakthrough in soil fertility at the time, and Joffe completed his Ph.D. thesis in 1922.

A few months later, Waksman wrote about these experiments in a scientific paper, putting his name first as the senior author. Joffe was shocked. He believed he was the one, not Waksman, who had discovered the microbe. Though he stayed on at Rutgers, becoming a professor at the college and an authority on soil science, Joffe nourished a dislike for Waksman, never ceasing to complain that, in his view, Waksman had stolen his work.

As an undergraduate, Schatz had taken a course with Joffe; he had gotten to know him well and had heard him complain that Waksman had not given him due credit. "Joffe would, not in class, but in his office, rant and rave about Waksman," Schatz later recalled. Some believed Joffe, others Waksman. The disagreement was never resolved, and the incident had left a question mark over Waksman's apparently amiable stewardship of the Department of Soil Microbiology.

Most of the time that he worked for Waksman, Schatz never gave the Joffe legend, and what he knew about it, a second thought. But now that he had his own discovery, he began to wonder how Waksman would respond. The two scientists had a close relationship at that point—partly,

Schatz was sure, because they were both Jewish, with roots in Russia. Waksman was not only his teacher but had also become a father figure, the male guide and mentor that he had not found in his own childhood growing up on a dirt farm in Connecticut.

He wrote up his results and gave them to Waksman to check and edit. The first scientific paper announcing streptomycin was published in the *Proceedings of the Society for Experimental Biology and Medicine* in January 1944. Schatz was thrilled. Waksman had acknowledged the crucial work Schatz had done by putting his name first, then Betty Bugie's. Waksman's name came last. It seemed that Schatz had no cause for concern that the Joffe case might be repeated, with him as the loser this time. But he was not privy to the behind-the-scenes struggle over the next stage of the discovery, the race to publish the effects of streptomycin on the deadly human strain of tuberculosis.

6 · The Race to Publish

AT THE MAYO CLINIC, WILLIAM FELDMAN received a copy of the three-and-a-half-page paper announcing streptomycin in February 1944. He and Hinshaw read down the list of twenty-two bacteria vulnerable to the new antibiotic and were astonished to find, among the usual test germs, *M. tuberculosis*. They were also amazed that there was no discussion anywhere in the text explaining the nature of the strain, whether it was the harmless kind or the deadly H37Rv. In reality, it was harmless, as Hinshaw would learn much later. Waksman told him that the strain of *Mycobacterium* used by Schatz was a non-pathogenic strain from the Department of Soil Microbiology collection. Waksman added that he would "not permit any pathogenic tuberculosis culture in his laboratory for fear of infection to students and technicians." But this rule was about to change—for Schatz's experiments in his basement lab. In reading the paper, Feldman also found it odd that Waksman had not mentioned this discovery when Feldman had visited Rutgers a few months earlier, though he must have known about it. The lead time on such journal articles was a month at least.

Feldman was about to contact Waksman when a letter came from the Rutgers professor himself, in which he offered to supply a sample of streptomycin to the Mayo laboratories for guinea pig trials. Feldman accepted immediately, telling Waksman he needed ten grams, which he estimated "from the meager knowledge available" would be enough for a small test—say, four to six guinea pigs. In exchange, Feldman agreed to send Waksman a culture of the virulent human strain, H37Rv, which Schatz needed for his in vitro (test tube) experiments.

In his basement lab, Schatz cranked up his small stills, frothy brews in glass flasks, to provide Feldman with ten grams. Schatz did not have to be told how to run a still; he had learned the tricks at an early age during Prohibition on the family farm in Connecticut. In the basement lab he now ran stills twenty-four hours a day, sleeping on the lab floor. He had an arrangement with the night watchman that if, on his rounds, he found Schatz asleep and the liquid in the flasks had evaporated below a red line, he was to wake him up. This punishing routine left Schatz permanently exhausted, and one morning he left the lab at about two o'clock and collapsed in the snow outside. Fortunately, the night watchman found him unconscious and called an ambulance to take him to the emergency room. He had pneumonia and spent a week in hospital. Waksman was the only member of the staff who didn't visit him.

Back at work after a week, Schatz began the crucial tests to see how his microbes would deal with the H37Rv strain. He took what precautions were available. The safety equipment had not progressed much beyond the crusted and cracked rubber aprons of the 1930s. By modern standards they were pitiful. He grew the germs in narrow test tubes, trying to limit exposure of the growing TB cultures to the open air when he removed the cotton wool caps. He had to sterilize all his own glassware. This meant putting the test tubes into the autoclave before leaving the lab—often in the early hours of the morning. The tests took much longer than those of the other germs because H37Rv was one of the slowest-growing—multiplying every two days instead of every twenty minutes like most bacteria. He started his first H37Rv test on March 24 and noted it as Experiment 10 in his new 1944 notebook.

It was a potentially deadly operation, handling germs that had killed millions of people without gloves or face masks or proper ventilation. The TB germ spreads on airborne droplets, but Doris Jones later remembered seeing Schatz wash out his mouth with antiseptic after a day's work. No other remedies were available. For the entire time he was using the virulent strain of TB, he later recalled, Waksman never came near the laboratory, and he told Schatz never to bring the germs to the upstairs labs. Schatz believed that his professor was afraid. It was not an unreasonable

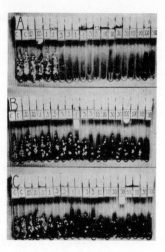

Rows of test tubes showing how streptomycin destroys the human strain of
tuberculosis, H37, in Albert Schatz's experiments in the basement laboratory. At left,
the tubes start with the TB microbes, which are gradually destroyed.
(Photograph by Julius Schatz, courtesy Vivian Schatz)

belief: In his retelling of the discovery of streptomycin, Waksman never
mentioned being in the basement laboratory during these crucial experiments.

By the middle of April, Schatz had made enough streptomycin to send
ten grams to the Mayo Clinic, and Feldman and Hinshaw started tests with
four guinea pigs on April 27, quickly running through the initial supply.
Schatz obliged by keeping his stills going night and day.

THE MAYO GUINEA pig experiment presented a major problem of logistics.
The animals had to be injected with streptomycin every five hours,
which meant Feldman and Hinshaw driving through the winter snows
to the lab, plus dealing with gas shortages and a lack of lab staff because
of the war. Feldman, with Waksman's permission, brought in a third
researcher, with "one of the finest labs," to test animals infected with
bacteria other than TB. Initial results were promising on everything but
cholera.

Waksman began to worry that he was somehow losing control of streptomycin. In the fiercely competitive world of patents and scientific discov-

ery, priority matters. The recognition and the rewards go to the scientist who publishes first. The research field was suddenly crowded. Schatz had done the initial isolation. Doris Jones had done the first toxicity tests on chick embryos. The vet at Rutgers was testing streptomycin on mice. Merck researchers were working on ways to extract the drug from the nutrient broth in which *A. griseus*, the streptomycin-producing microbe, was grown, and the company had also done animal tests. And besides TB, the Mayo Clinic was also testing streptomycin on plague, tularemia, pneumonia, and cholera. Understandably, Waksman wanted his results—or, more correctly, Schatz's results—on in vitro experiments with H37Rv to be published first. The Mayo researchers had a quick and easy route to publication through their own *Proceedings of the Mayo Clinic*, which was published every two weeks. The publishing routes for Waksman took a month at least—as Feldman had surmised.

Waksman told Feldman that he wanted to be sure that the Mayo people would warn him if they were going to publish anything, and would let him "look over critically" any papers reporting results from their experiments. This would avoid "any possible confusion" among the various research groups. Feldman agreed.

Feldman and Hinshaw ended their initial four-guinea-pig test on June 20 after fifty-four days. Streptomycin worked against TB, and it was not toxic. They concluded, "The new substance was well tolerated at a therapeutic level sufficient to exert a marked suppressive effect on otherwise irreversible tuberculous infection in guinea pigs." And after five years of trying, this was the most positive result they had ever seen; perhaps they had even found a cure.

There was one important qualification. Like the sulfa drugs, streptomycin had not destroyed all the TB germs, and therefore the action of the drug was bacteriostatic—it stopped the infection from spreading—but not bactericidal, meaning it did not kill the germs. How much streptomycin was an advance on the sulfa drugs was not yet clear. There could also have been errors, but the two researchers were confident of at least one part of the trial: Their tests had shown no adverse toxicity.

Feldman and Hinshaw celebrated their findings in the shade of an elderly apple tree in Feldman's garden. They sipped one of Feldman's "mysterious and delicious concoctions made with laboratory alcohol [200 proof]." But they kept reminding themselves that this was but "one step

William Feldman (left) and Corwin Hinshaw (right) of the Mayo Clinic in
Rochester, Minnesota. They confirmed streptomycin's effect on tuberculosis
in guinea pigs and later carried out the first clinical tests on humans.
(Feldman by permission of the Mayo Clinic Historical Unit, Rochester,
Minnesota. Hinshaw courtesy of Horton Hinshaw and Dorothy Hinshaw Patent.)

forward" in their search. The important next step was to try streptomycin
on a larger number of guinea pigs over a longer period of time.

Before the sun had set in Feldman's garden on that first evening, they
had decided on a plan. To make a more certain appraisal, they needed to
give the slow-growing TB germ time to take a proper hold in the guinea
pigs' organs before starting the streptomycin treatment—and this meant
infecting the animals six to seven weeks before the start of therapy.

They also saw a problem of priority. Waksman had indicated that other
labs were involved, but he had not named them. Hinshaw, always impul-
sive, suggested that they start to infect guinea pigs immediately. Although
this was strictly against the Mayo Clinic rules—the clinic first had to give
its approval—Feldman agreed. He was as impatient as his colleague.

At the same time, Feldman arranged a meeting with Waksman to collect more streptomycin, but did not mention that they had already started
the second experiment. It was a gamble, of course. Feldman thought his
request for more streptomycin would carry more weight if Hinshaw came
to the meeting, as he was eager to do, and he did not give Waksman a
chance to say no. He made reservations for two at the Roger Smith Hotel

in New Brunswick for the night of July 9. "We will contact you sometime Monday morning," he wrote to Waksman.

Hinshaw was shocked at the meager facilities of Waksman's lab—and even more surprised that Waksman had never worked with animals himself. He was also curious to find that Waksman had no interest in pathology and no medical expertise. He was a soil scientist, and his labs were not set up to produce the quantities of streptomycin needed. "It was obvious that he would not be able to produce the streptomycin we wanted for an adequate experiment," Hinshaw concluded. Schatz had exhausted his supplies, and in any case his primitive production line was simply not up to the task.

Waksman arranged for Feldman and Hinshaw to meet Merck officials to see if some of the company's own sparse supplies could be released. Waksman asked them to keep this contact with Merck secret; he didn't want other researchers, who were also interested in samples of the new drug, to know that Merck could release supplies—if, indeed, Merck decided to do so.

When they arrived at Rahway, the Mayo team were surprised to find more than a dozen Merck researchers gathered for the "secret" meeting and assumed they must be in the wrong room. They quickly retreated and continued down the corridor. But they were called back. They presented their "meager evidence," and the Merck chemists at first said no, they could not increase their production. Feldman and Hinshaw were about to give up when George Merck himself walked into the meeting. Hinshaw happened to be describing the prevalence of TB among World War One veterans, and Merck nodded his head. He could justify providing the Mayo Clinic with scarce streptomycin as part of the war effort.

In the end, they got their streptomycin, but as they left, Waksman warned them again against any "premature publicity." Other companies also researching the new drug would be upset if they knew, he stressed. The Merck supply should be "privileged information." In fact, Feldman and Hinshaw found his warning unnecessary. In their labs, they had always been careful never to use the real names of drugs that they were testing, referring to them only by code names.

With new supplies now assured, Feldman and Hinshaw began two crucial experiments. The first was to run over thirty days—"to determine if the initial experiment could be confirmed." It was important to know if the previous experiments could be repeated. The second, a longer-term experiment, would last into January 1945.

At the same time, the results from the third Mayo researcher, Dr. Fordyce Heilman, were also positive. All five of the harmful bacteria, including those causing plague and bovine TB in infected mice, were "completely inhibited" by streptomycin. The results were "considerably better" than those demonstrated in experiments using sulfa drugs and the pressure was building on Waksman to let the Mayo team publish.

IN THE MIDDLE of September 1944, the thirty-day guinea pig test at the Mayo Clinic proved to be a "definite" confirmation of the earlier success. Feldman wanted Waksman to be the first to know and telephoned him with the news, arranging for his secretary to record the historic call in shorthand. This experiment had used two virulent strains of TB, H37Rv and a sputum strain, which he and Hinshaw had isolated a year earlier. Feldman, a cautious man, used the word "definite" to describe the results, but it was underplaying a dramatic breakthrough in the long fight against TB. Oddly, Waksman was unmoved. He was preoccupied, distant.

"Good," he said, according to the transcript.

Feldman pressed on with the astonishing news. To a man of few exclamations himself, this seemed a minimal enthusiasm, at best.

"There can be no reasonable doubt that streptomycin has a considerable efficacy in combating experimental TB."

"Fine," Waksman responded.

"I will write you a more detailed account of the results, probably tomorrow," Feldman said.

Waksman: "Good. I shall look forward to hearing of the results."

Waksman's seeming lack of interest was not out of character. He was an intensely practical man, not known for eloquent conversation, especially about work. He had a fiercely logical mind, moved swiftly step-by-step, and did not like to waste time. But what Feldman spoke of next suddenly got his attention: publication of the Mayo results.

"We would like to know your wishes regarding the publication within the next month or so of a preliminary report of this work."

Waksman hesitated.

"Well," he said finally, "I think it would be all right. I'll tell you what I will do. We have prepared for publication our in vitro study [Schatz's streptomycin tests on H37Rv], for the *Proceedings of the Society for Experimental Biology and Medicine*. I don't see why you couldn't do that for your in vivo [guinea pig] study."

This was not what Feldman had in mind. The society's journal would take a couple of months at least to publish Schatz's results; the Mayo Clinic's own publication could print their results in two weeks. So he ignored Waksman's suggestion, continuing, "We would of course submit to you and the Merck people a copy of the manuscript for suggestions before submitting the paper for publication."

Waksman summoned an ally. He said that he personally saw "no objection" to the Mayo publication, but he wasn't sure "how the Merck people will feel." Merck was also doing animal tests. But Waksman's aim was to stall Feldman until he could publish Schatz's crucial in vitro tests. That paper would also have Waksman's name on it, of course, even though Schatz had done the experiments alone in the basement lab. Waksman wanted to leave Feldman with the understanding that there might be a doubt about Merck's continuing to supply streptomycin if he didn't comply with Waksman's wishes and delay publication.

Feldman understood the warning. He could not afford to offend Merck. Even though ten companies were now working on the drug at various stages of production, they were about six months behind Merck. There was no alternative supply.

Even so, as soon as he put the phone down, Feldman wrote Waksman, pressing him again to let the Mayo Clinic publish its results. Feldman felt "it would be proper" to prepare a brief report of his observations for publication. This would be only a preliminary notice to establish the Mayo team's priority in the animal tests, and would be followed by a much more detailed account at a later time.

"Since you indicated that you had no objections to the publication of a brief report, I will determine that the people at Merck's do not object . . . We will of course send for your information a copy of the manuscript."

Waksman did not reply directly; he passed his problem of priority on to Randolph Major, the director of research at Merck. In a gracious but firm letter, Major told Feldman to delay publication. He "quite understood" Feldman's desire "to publish these very gratifying results," but he had talked it over with Waksman, who "felt that you might care to wait until the publication of his initial in vitro results."

Over Hinshaw's objections, Feldman agreed to a delay. He had "discussed very frankly the entire situation" with Hinshaw, he told Major. "We are in entire agreement with Dr. Waksman's wishes that we withhold publication until such time as his report on the in vitro studies is published." But he hoped this would delay would probably not exceed four to six weeks.

Hinshaw was upset: He wanted his and Feldman's work to be acknowledged as soon as possible. This was a breakthrough for them—after five long years of trials. They had completed their initial experiment with four guinea pigs *before* Schatz had completed his in vitro tests. He didn't see why they had to wait.

In the end, Waksman won the day, overwhelmingly. He established priority not only over Feldman and Hinshaw but also over Albert Schatz. The Mayo Clinic invited Waksman to give a lecture at its Rochester, Minnesota, headquarters on October 5, 1944. It was written up in the Mayo's *Proceedings*, and Schatz's results beat Feldman and Hinshaw's into print.

On the day of the lecture, the Mayo Clinic director, Dr. D. C. Balfour, introduced Waksman. This is how Waksman later wrote up Dr. Balfour's remarks.

"We have with us today a representative from an agricultural institution, one who has no medical degree and has never received any medical training. He is going to address us tonight on a subject that is at the moment of great importance to medical science and clinical practice, dealing with the discovery and application of new chemotherapeutic agents in the treatment of infectious diseases. The fact that you, Dr. Waksman, have been invited to deliver an address before a great medical organization, such as ours, suggests that you are bringing to us a very important message."

Then Waksman told the assembled physicians of his background on the Rutgers college farm, of his longtime interest in the actinomycetes, and of his transition from compost and humus to antibiotics. Two events in 1939, he said, "provided the necessary stimulus." One was a former student's work on gramicidin—he did not mention René Dubos by name. The other was World War Two and the need for new agents.

He recalled finding the first antibiotic, actinomycin, but determining that it was very toxic. The second, streptothricin, was "a great step forward," but it was also toxic. Finally, "in September, 1943, *my assistants and I* [emphasis added] succeeded in isolating in our laboratory an antibiotic which possessed properties similar to those of streptothricin but was less toxic." This was streptomycin.

Albert Schatz was not at the lecture, and Waksman did not mention his name. Feldman and Hinshaw were in the audience, but in Waksman's retelling of the events, they were not mentioned either. Waksman gave equal importance to streptothricin and streptomycin, even though he had already been informed by Merck that, in its opinion, streptothricin was too toxic for use as an injected drug.

His paper was published on November 15 by the Mayo Clinic, and this time his name came first and Schatz's last, with Betty Bugie's in the middle. Schatz's prospective paper on his in vitro work on H37Rv, where his name would be first and Waksman's second, was mentioned in a footnote as being "in press." Feldman and Hinshaw's work was not mentioned. In addition, Waksman created a myth about how Schatz had isolated his streptomycin-producing strain of *A. griseus.* He told the Mayo physicians that antibiotics in his lab were isolated according to a "six-step" method that, he implied, he had devised. He wanted to leave the impression that he should be given credit for the way Schatz had isolated his microbe. The first two of the six steps were ways of finding which bacteria were likely to produce antibiotics, including soil enrichment. (The other four steps, including the streak test, involved testing the microbes, once found, to see which germs they were capable of destroying and then how to produce the antibiotic.)

In Waksman's Mayo Clinic paper, he said that "both streptothricin and streptomycin were isolated and studied by the use of these procedures." This was not true, and as the supervisor of Schatz's thesis, and one of his examiners, he should have known that. The thesis shows that Albert Schatz did not use the first step in isolating his microbe. And when he used the second step, he got no positive results. If he had used only Waksman's "six steps," the chances are that he "would not have isolated" *A. griseus* because those first two steps did not produce a clear zone. He isolated the streptomycin-producing strain through random selection, using the same kind of intuition that a gardener might use to pick a good, healthy plant for breeding. However, Waksman's myth would create the impression among many observers of his antibiotic program that all his graduate students, including Schatz, followed a strict method, laid out before Schatz started his Ph.D. That Schatz found his microbes through the serendipitous method of random selection and that "perhaps some intuition was needed" would not be acknowledged by Schatz's peers for another half century.

THAT EVENING AT the Mayo Clinic, Waksman left a manuscript of Schatz's in vitro paper with Feldman, who, again prodded by Hinshaw, asked for at least a footnote on their own work to be included. "It would very properly give you the credit for recognizing the importance of the in vivo studies before speculating as to possible chemotherapeutic efficacy," he said. Such a footnote, he continued, might read, "At the suggestion of the authors the effect of in vivo tests of streptomycin on *M. tuberculosis* is now being studied by Feldman, Hinshaw and Heilman, Mayo Foundation, Rochester, Minnesota. The results of their study will be the subject of an early report." The footnote was printed in the paper, but without the last sentence. Waksman had successfully concluded his first move to control the story of the discovery of streptomycin.

7 · A Conflict of Interest

BY THE SPRING OF 1944, THE American companies producing penicillin were making enough to meet Allied demand. From D-day,
June 6, 1944, onward, the death rate from infected wounds was reduced dramatically, and by 1945 penicillin was available for civilians. As
streptomycin continued to show promise during the summer of 1944, Dr.
Waksman was rarely seen in the "salt mines," as Doris Jones had christened the laboratories where the graduate students toiled, poring over their
petri dishes for zones of antagonism. Waksman's onetime fascination with
new species of mold, with watching them grow and perform weird tasks
against their cousins, had long since faded. He was now preoccupied with
turning streptomycin into a new wonder drug. In an astonishing move,
Merck would give up its exclusive right to produce streptomycin—the right
that Waksman had negotiated in return for Merck's giving three hundred
dollars a month to his department. Under a new arrangement, the patent
rights would be owned by the nonprofit Rutgers Endowment Foundation,
which would license companies to produce the drug, paying the foundation a 2.5 percent royalty. Waksman would be in charge of arranging those
licenses. He would become the overlord of streptomycin.

Any thoughts he might have had about leaving Rutgers for a better job
at a bigger university, or even giving up academic life altogether and going into industry, quickly disappeared. He would stay at Rutgers, despite
the tiny size of his department, despite Rutgers's lower academic standing
when compared with larger "aggie" colleges such as Cornell or Wisconsin,

and despite the school's meager level of funding. At fifty-six, Waksman was about to have his world transformed—and so were the drug companies.

The official reason for Merck giving up the patent was that Merck and Rutgers, realizing how significant the new drug could be, had agreed that no one company should have a monopoly. Streptomycin should be produced by as many companies as possible, just like penicillin. More would be produced at a cheaper price because of the competition. Merck was seen as magnanimous in relinquishing what was about to be a grand money-spinner, and Waksman was seen as a great humanitarian for persuading the company to take this extraordinary path. While there was some truth in these official explanations, the reality, as with so many things unfolding in the streptomycin story, was somewhat different.

SINCE THE BEGINNING of the war, the Allies had become increasingly concerned about the enemy's possible use of biological weapons. Among the secret reports was one concerning a bizarre prewar operation in February 1939. Japanese agents tried to buy yellow fever virus at the Rockefeller Institute, in New York, but the plot was uncovered by the Federal Bureau of Investigations. In October 1941, the secretary of state for war, Henry Stimson, had called on the National Academy of Sciences to form a committee to assess the current state of knowledge about biological warfare. A "war bureau of consultants," a dozen scientists led by Dr. Edwin Fred, professor of bacteriology at the University of Wisconsin, assessed the enemy's possible offensive and defensive arsenals. In May 1942, President Roosevelt authorized Secretary Stimson to establish a secret civilian agency, the War Research Service, to develop biological weapons, which the United States would keep in reserve and never be the first to use. A site was chosen at Fort Detrick, in Maryland, an old army base. The War Research Service was made part of the Federal Security Agency to obscure its existence, and George Merck, who had proved his patriotism with his "Command me" remark to Vannevar Bush, was named director.

In 1942, intelligence reports on enemy development of biological weapons escalated. An FBI informant in Tokyo had overheard drunken German doctors saying they were teaching the Japanese about biological weapons, including anthrax and typhus. Another informant, in Los Angeles, warned that Japanese saboteurs had cans of a jellylike substance containing typhus and bubonic plague. No such containers were ever found. A Swiss report said that the Germans would use "bacteria of every description" on Allied forces. Yet another said that Germans were making biological weapons in laboratories in Brazil. In March 1943, a memo from the War Department on the enemy's "resort" to biological weapons talked about possible contamination of the U.S. domestic food supply.

Although U.S. policy on biological weapons was strictly defensive, that did not prohibit the Office of Strategic Services, the forerunner of the Central Intelligence Agency, from asking Merck's War Research Service to recommend "three or more" of its "most potent organisms" for use in covert operations, with directions on how to use them. At a special meeting chaired by George Merck to deal with the OSS request, the experts, led by Dr. Fred, suggested "two types" of organisms—"those extremely pathogenic and those that reduce the efficiency, but may not be fatal." The "most promising" were "B. anthrax and Cl. botulinum," and "perhaps a dysentery organism of the *Shiga* type, the plague organism [and] brucellosis of the *Suis* strain should be studied." Streptomycin had shown promise against the plague and brucellosis. Later the OSS would ask for pellets or capsules of *Staphylococcus aureus* for "definite war use." Albert Schatz had used *S. aureus* as one of his test organisms in his hunt for streptomycin.

GEORGE MERCK HAD a clear conflict of interest. Under the deal with Waksman, his company would have monopoly control over streptomycin, which, he knew, had already proved itself to be effective in fighting the kinds of diseases that might be spread by the enemy in a biological weapons attack. And he was the chairman of the president's committee charged with finding such a drug and producing it as quickly and as cheaply as possible. In addition, Merck was under constant pressure to increase supplies of streptomycin, and it had experienced production problems. At one point, they lost the entire contents of three of the big thirty-thousand-gallon fermentation tanks producing streptomycin because of a virus invasion that killed the culture. George Merck took the only path open to him: He gave up the company's monopoly rights—but on his terms. Dr. Waksman understood Merck's position very well. Since 1942, he had been advising Merck's committee on how to deal with fungus infestations of clothing in the tropical climates where U.S. forces were fighting. And he was prepared for this moment.

In October 1943, after Schatz became convinced of his discovery of streptomycin, Waksman had informed Merck's Randolph Major that he had received a dramatic British report about the treatment of burns. "We can keep most of the burns uninfected for the first 10 days or so (by the use of sulphonamides and penicillin)," the report said, but to defeat subsequent infections, it continued, the British doctors needed a new antibiotic active against Gram-negative organisms.

At the time, Waksman still hoped Merck would be able to turn strepto-thricin, discovered by Waksman and Boyd Woodruff in 1942, into that much-needed drug. Under his contract with Merck, Waksman had agreed not to give out samples of streptothricin to any other companies for testing. But he asked Merck to waive this rule.

"It is hardly fair on my part not to place a material that we have isolated and that proved to be active against Gram-negative bacteria, especially a non-toxic preparation, in the hands of those that need it," Waksman wrote. He still hoped that somehow it could be detoxified.

Major "quite understood" Waksman's position. He agreed to release the cultures "if you find it necessary." The only reason Merck had felt justified in asking Waksman to withhold these cultures, he said, "was that I understood that you did not object to doing this."

This was evidently the beginning of several conversations and letters between Waksman and Merck during which Waksman told the company of his "feeling" that he should be "free to make available for development by other organizations antibiotic agents which may be of value to the armed forces in the present emergency."

By the end of 1943, the OSS had reported again that the Germans might be planning a biological weapons attack. While the evidence was still inconclusive, secret defensive work on biological weapons in the United States, Britain, and Canada was producing "concrete information" that such an attack "was feasible." In the New Year, "all work in this field" was stepped up, so that by the summer of 1944 a large part of the program had been handed over to the War Department. President Roosevelt established a new overall manager, the U.S. Biological Warfare Committee. Merck was appointed chairman. His conflict of interest remained and, if anything, was heightened.

As "consultant to the secretary of war," Merck was personally receiving intelligence reports on enemy biological warfare. One such OSS report in August 1944 claimed that the German army had a "special service" in occupied France that was sending vials of cattle plague to England to be used for spreading epidemics among British livestock. The German agents in Britain were supposedly to infect the cattle when they were brought to market. In planning how to retaliate "in kind" if necessary, a "wide variety of agents pathogenic to men, animals and plants was considered." This included many bacterial agents, such as bubonic plague and tularemia, against which streptomycin was the latest, and most effective, drug.

The War Production Board, now in control of U.S. industry, considered bringing streptomycin production under government control, as had been done with penicillin. But such a move would have required a special act of Congress. Unlike with penicillin, no government research funds had been allocated to the acquisition or distribution of streptomycin. Thus the government owned no rights to it. The drug had not even been cleared or approved by the Food and Drug Administration and therefore could not be sold publicly for therapeutic purposes, "except to the federal services"— i.e., the military, including the biological weapons program, or government agencies performing clinical trials.

In June 1944, Waksman formally asked Merck for the agreement of November 8, 1940, giving exclusive drug-development rights to Merck to be "abrogated." As Feldman and Hinshaw at the Mayo Clinic announced the first successful guinea pig trials, George Merck ordered his legal staff to draw up a letter to Waksman ending their deal. The draft referred to unspecified "correspondence and conversations" with Waksman.

On August 17, 1944, this draft became a formal letter and was signed by Waksman and William Martin, dean of the College of Agriculture, and eventually by Rutgers president Robert Clothier. The final letter kept the same wording: "You have advised us of your feeling that you should be free to make available by [sic] other organizations antibiotic agents which may be of value to the armed forces in the present emergency."

Later the agreement would be attributed by Rutgers to the fact that Merck and Waksman had realized that their cooperative enterprise had resulted in a great humanitarian discovery which should not be subject to an exclusive license. Accordingly, Merck & Co. Inc. had "voluntarily abandoned" its patent.

The return of the patent did not come free, as the fine print revealed. The company only agreed providing it could have a nonexclusive license and a new agreement "satisfactory to us." In the new agreement, Merck demanded repayment of the estimated $750,000 worth of research it claimed to have carried out for Waksman on development work on streptothricin and streptomycin. This was mostly work done by the company's chemists in extracting and purifying the drugs. Rutgers eventually agreed to $500,000, which would be paid back to the company from Rutgers' future royalty earnings. Also, as part of the new deal, Merck agreed to continue to use its legal staff to make the patent application.

ON AUGUST 14, 1944, Waksman and Schatz filled out a standard memorandum of invention titled "Streptomycin and process for producing." The date of the "conception" of the discovery was listed as August 23, 1943, the date of Experiment 11 on page 32 of Schatz's notebook. The first verbal disclosure was on September 10, 1943, when Schatz told Waksman's assistant, Robert Starkey. Waksman was apparently away. The first written description was recorded as October 8, 1943, and in Waksman's notebook as his Experiment 59.

The date of "reduction to practice" was put as November 22, 1943, corresponding to Experiment 72, on December 30, 1943, in Waksman's lab notebook. That experiment was titled "Influence of treatment on the extraction of streptomycin." On the opposite page, Waksman wrote in later, "First time method of extraction of streptomycin is described in detail." Thus, according to the patent application, Waksman and Schatz were officially "co-discoverers."

The form was signed at Rutgers with Merck lawyers present. For the first time in their close relationship, Schatz questioned Waksman. He asked why they had to have a patent and why they had to use Merck lawyers. The university was an independent institution; why did they have to be involved with a commercial concern? For the first time, Schatz stood his ground with Waksman. "Streptomycin was the fruit of my labors. I felt that anything pertaining to human health and human life should be made available as quickly and as cheaply as possible to all people . . . I didn't want to have anything to do with Merck, whose interests were profits," he recalled telling Waksman.

Waksman replied that they had to go ahead with the patent to protect streptomycin. Without a patent, a company like Merck, Squibb, or Pfizer might produce a derivative of the drug, take out a patent, and control production and prices. Merck had patent lawyers who were able to help Waksman and Schatz with the application, a service that would otherwise have been expensive and was something Rutgers could not afford. Schatz would later remark, "Until then, the thought of patenting had never entered my head."

BEFORE THE YEAR was out, the first publicly recorded tests on human patients were carried out in New York and at the Mayo Clinic. Both tests were successful.

On September 21, 1944, a two-week-old infant arrived at Columbia University Babies Hospital, in New York, suffering from a bacterial infection that had caused meningitis, septicemia, and a "heavy urinary tract infection." The infant was deeply jaundiced, and his liver and gallbladder were enlarged. The doctors gave him a sulfa drug, sulfadiazine, combined with penicillin, but the infant's fever remained high, and he was clearly dying.

In desperation, but with no clinical data to support their decision, the doctors switched to streptomycin, administering it every three hours for five days, then doubling the dose on the last, the sixth, day. The infant's temperature suddenly dropped, and he recovered. A memo of the event stamped "Secret: not to be disclosed without special permission" appears in Dr. Waksman's archives. One of the doctors wrote, "The medical staff at Babies Hospital are naturally very much interested and excited about this case. They have no doubt that streptomycin produced the favorable change in the clinical procedure." There is no record of whether Waksman was involved or who had provided the streptomycin. It could have come directly from Merck and Waksman may not have been involved. Merck was testing streptomycin on mice, rats, guinea pigs—and humans. Merck had also tested streptomycin on six patients suffering from Gram-negative bacterial infections. The results were mixed. An adult woman with endocarditis—infection of the heart lining—and an infant with sepsis died after treatment. The antibiotic had no effect on a child with tuberculosis and meningitis, or on another with brucellosis. An adult with typhoid fever recovered. The drug had promising results on a patient with pneumonia.

On November 20, 1944, Patricia, a twenty-two-year-old farmer's daughter, lay dying in a hospital bed at the Mineral Springs Sanatorium, in Cannon Falls, Minnesota. She was suffering from advanced and spreading pulmonary tuberculosis, the most common kind. Patricia had been in the sanatorium for more than a year under the care of Dr. Karl Pfuetze, the medical superintendent of the sanatorium, and his assistant Dr. Marjorie Pyle. They had treated Patricia with the conventional forced bed rest. After initial improvement, she had deteriorated. Her right lung was badly diseased, and she suffered from alternating chills and high fever, sweating, and a worsening cough. The doctors considered her to be near death, and recommended that she be transferred to the Mayo Clinic for an assessment by Dr. Hinshaw.

With Patricia's consent, Hinshaw immediately started a course of streptomycin injections, using his limited and impure supply. He could only guess at the proper dose, and he was cautious at first, steadily increasing the dose until Patricia showed signs of improvement. The infection, which had spread to her left lung, gradually disappeared. Eventually, the doctors were able to operate on the right lung, removing the diseased section. Patricia made a remarkable recovery and was released from the sanatorium. She eventually married and had three children.

Within a week, in collaboration with Pfuetze, Hinshaw started to give streptomycin to twenty-two patients from the Mineral Springs Sanatorium; eighteen improved. These were Hinshaw's "scouting phases," and he played down the results. He emphasized that many forms of TB "tend to improve spontaneously without treatment," and that no single drug was likely to have "a rapidly curative effect" in a disease with the clinical pathology of TB. He did "not expect improvement" in less than a few weeks, nor did he expect clinical arrest of the infection in less than a few months. He pleaded for "extreme conservatism" in judging clinical results regarding chemotherapy and TB.

Feldman and Hinshaw refused to talk to newspapers about the results unless they were allowed to check and edit the story. Reporters also had to clear their copy with the Minnesota State Medical Association's Committee on Tuberculosis. The newspapers cooperated. Feldman and Hinshaw were equally cautious in articles published in medical journals, avoiding overly optimistic statements that might mislead the medical profession. They pointed out that extensive and prolonged clinical investigation would be required to determine the place of streptomycin in the treatment of TB.

On January 20, 1945, however, the successful end of the third and largest guinea pig trial was a real cause for celebration. Feldman cabled Waksman, "Long term crucial experiment streptomycin terminated today. Incomplete results indicate impressive therapeutic effects." In a follow-up letter, Hinshaw wrote, "The results are sufficiently encouraging to be tantalizing . . . If we could give a million or more units a day we might have something more impressive."

Also at the end of January the Mayo researchers completed the first clinical trials on patients from the Mineral Springs Sanatorium. They reported a total of fifty-four cases of TB in which the patients had received streptomycin for a period in excess of four weeks, and the number of cases had increased to seventy-five by June. For the first time, they found some toxicity affecting the eighth cranial nerve, resulting in some dizziness. It was a small note of caution in an otherwise optimistic report.

The U.S. government prepared to set up a distribution network similar to the one used for penicillin, so that limited supplies would go to the army first. Merck sent small quantities to the Army Medical Corps, the U.S. biological warfare program at Fort Detrick, and the British chemical and biological warfare establishment at Porton Down, in Wiltshire, where they were testing streptomycin against an array of toxins. The official view was still cautious. Norman Kirk, the U.S. surgeon general, warned that "no conclusive statements" could yet be made as to the drug's potential because it was in such short supply.

A team of fifty Merck scientists was assigned at once to do everything possible to transform streptomycin from an extremely promising experiment into a therapeutic agent ready for use by doctors around the world. Merck broke ground on a $3.5 million plant, slated to employ four hundred workers, in Elkton, Virginia. On February 8, 1945, Merck lawyers filed the patent application papers of Selman Waksman and Albert Schatz for "Streptomycin and Process of Preparation." The application included an oath, sworn by Waksman and Schatz, that "they verily believed themselves to be the original, first and joint inventors" of streptomycin, plus an affidavit of Waksman's describing streptomycin as "the new antibiotic that Schatz and I have discovered."

PART II · The Rift

8 · The Lilac Gardens

AT THE BEGINNING OF 1945, AS the Allied armies in Europe prepared for the final push to Berlin, Albert Schatz was about to turn twenty-five. He planned to mark his surprising yet spectacular contribution to the war effort with a rare personal celebration. He was going to marry Vivian Rosenfeld, a bright, pretty, blue-eyed student with long dark curly hair who was studying biology at the New Jersey College for Women, on the Rutgers campus.

He had met her by chance a year earlier. He was spending such long hours in the basement laboratory that he rarely had time to make friends with eligible young women, even though the Women's College was a five-minute walk from his laboratory. Biology students like Vivian often saw Albert in his white lab coat when they came to mycology lectures in the Administration Building. They noticed his good looks, and they heard from the other graduates about his brilliance as a researcher, but he was always working. For his part, Schatz reckoned that even if he asked any of them out, they would not really have time for him. He lived in the greenhouse and had no money to spare at the end of the month, hardly ten cents for an ice cream. Certainly he could not afford a movie. The best he could offer was a stroll across the college farmland.

In the spring of 1944, however, he found one young student who liked to walk with him. She lived in one of the campus dormitories of the Women's College. One evening after work, he telephoned her from the lab, but Vivian Rosenfeld answered the phone instead. The student Albert had called for was out, but Vivian said yes, she would love to go walking, if he'd accept her as a substitute. Schatz agreed, and they set off for the Lilac Gardens, an area of woodland reserved for ornamental plants about two miles beyond Poultry Pathology and the milking shed.

That evening began a friendship, and soon followed love, which lasted for the next six decades. They swapped family histories. Vivian's grandparents were refugees from Ukraine, like Waksman. They had arrived in Philadelphia with no knowledge of English. Albert's back problems for once were not bothering him, and he made Vivian laugh by hanging on to a post and stretching his legs horizontal to the ground. And Vivian impressed him with her determined strides; fit as he was now, he had trouble keeping up with her.

It was the first of many walks around the farm, each season providing its own natural attractions. In summer they looked for wildflowers, and in the fall they picked mushrooms. They were especially drawn to the slime molds, the unfortunate nickname for a gelatinous microbe, in unusual browns and yellows, commonly found attached to the underside of deciduous logs on the forest floor.

As the friendship developed, Vivian would come over to the basement laboratory after hours and knock on the window. Albert would let her in, and she would do her homework while he attended to his experiments,

Albert and Vivian on one of their walks in the Lilac Gardens at Rutgers.
(Courtesy Vivian Schatz)

producing crude extracts of streptomycin. On Saturday nights in winter, when it was too cold to go walking, they stayed in the basement lab, going through Albert's slides, identifying various species of fungi and bacteria. It was an odd courtship, but it suited them perfectly.

The couple married on March 23, 1945, in a synagogue in Passaic, and left immediately for their honeymoon in Connecticut. Albert wanted to show Vivian the farm where he had spent his childhood, near Norwich. It was the first week's holiday he had taken since he had come back from the army two years earlier, and he couldn't leave his work and Dr. Waksman behind. In his shirt pocket were four test tubes with white cotton wool stoppers containing *A. lavendulae*, the bacterium that produces streptothricin. A week was too long to leave them unobserved, he explained, and Vivian wondered whether there had ever been another man who took test tubes of multiplying microbes on his honeymoon.

One morning, without telling Vivian, Albert even took time to write a letter to Waksman. He and Vivian had found the largest bracket fungus he had ever seen and carried it three miles back to the hotel, he wrote. "Each morning and night Vivian and I examine the four agar slants of the different colony isolates of *A. lavendulae*. Vivian says it's strange to have 'business' with us now, but she is as interested in the cultures as I am." Surely, this was true love.

BACK AT WORK the next week, they moved into a small apartment that they shared with another graduate student, and Schatz finished his thesis. Titled "Streptomycin: An Antibiotic Agent Produced by *Actinomyces Griseus*," it was approved for a doctorate on June 15—two years to the day after he had been discharged from the army. The normal residency requirement of three years was waived. The war in Europe was over, but lack of funds and shortages of everyday supplies at Rutgers meant that the 127-page thesis in which he had described each step of the discovery of streptomycin was not printed. The only copies available were carbons from the department's typist. Schatz signed one "To Uncle Joe from Albert," for his mother's brother.

The acknowledgments were generous. As he was bound to do, he thanked everybody involved, starting with "Dr. S. A. Waksman for suggesting the problem investigated and for his close supervision and encouragement throughout the course of this work." He also thanked Robert Starkey and Walton Geiger, the department's chemist, "for their interest and helpful advice," and his colleagues in the department for their "friendly cooperation."

The text mentioned that one of the strains had come from the swab of a chicken's throat, and Schatz's notebook made two references to the culture's having come from Doris Jones. He had written Jones into the history books. A final note said, "For the sake of completeness in the treatment of subject matter, a few experiments carried out in collaboration with Miss Elizabeth Bugie and Miss H. Christine Reilly have been included." Waksman was, of course, one of the examiners of his thesis, but there is no record of his making any comment.

Schatz thought that his thesis was the document which the outside world would rely on for evidence that he was the one who had actually discovered streptomycin, and he thought that his name coming first on the two key scientific papers written up from his thesis would support that claim, but he was in for a shock.

While Schatz was working in his basement laboratory, Dr. Waksman was upstairs in his office writing up his own account of the discovery. In March 1945, he published his first book on microbe wars, a masterly 350-page compilation of scientific papers going back to the first 1890 experiments with actinomycetes and forward to Schatz's discovery of streptomycin. The book, *Microbial Antagonisms and Antibiotic Substances*, described Schatz's discovery as follows: "Certain strains of *Streptomyces griseus* produce an antibiotic substance, designated streptomycin, that is also active against both Gram-positive and Gram-negative bacteria." Because of its low toxicity it had "great promise of practical application." (In 1943, Waksman and a colleague reclassified the actinomycetes. *A. griseus* was put into a new genera, Streptomyces. Thus, Schatz's *A. griseus* may now appear as *S. griseus*, depending on the original text.) The name Albert Schatz did not appear in the book, only in the bibliography.

With the book's publication, Waksman began to give interviews to the popular media, portraying himself as the "discoverer" of streptomycin and Schatz as "one of his assistants." A breathless account in *Liberty Magazine* titled "Keep Your Eye on Streptomycin" was a "preliminary report on what is potentially the biggest medical news since penicillin ... Streptomycin appears to open the way to the conquest of *all* infectious diseases ...

Potentially, swift cures for everything from watermelon wilt to infantile paralysis lie hidden in the grubby soil."

Streptomycin was also said to be effective against undulant fever, which under the name of Bang's disease cost cattle breeders $30 million a year. Nearly two thousand cases a year in the United States of the some-times fatal tularemia might be cured with streptomycin. There appeared to be an excellent chance that streptomycin would be an unparalleled weapon against plague and leprosy, hog cholera and Dutch elm disease. But "its greatest, most exhilarating accomplishment is its action against the tubercle bacillus, which causes tuberculosis."

IF SCHATZ SUBMITTED to this loss of stature, his parents and especially his mother's brother, Uncle Joe, were outraged to find him relegated to the level of an assistant to the "discoverer," Waksman. Uncle Joe had recently qualified as a dentist and, being of the kind known disparagingly in the profession as "an advertising dentist," understood a thing or two about publicity. Although he never took credit for what happened next, the front-page headline on the July 2 issue of the Schatz family's local paper, the *Passaic Herald-News*, declared, PASSAIC YOUTH DISCOVERS DRUG THAT MAY STAMP OUT DREAD TB. A reporter had visited the Schatz family at home and found "a slim youth of 25, a product of Passaic's public schools and now engaged in research in soil microbiology," who was "on his way toward startling the world of medical science with a drug more wonderful than penicillin." For Schatz "and those associated with him" it had meant "hours, days, and months of tedious, painstaking and oft-times discouraging lab-oratory work."

The entire family had gathered for the interview. Albert's father, Julius, a housepainter and part-time dirt farmer, was "proud" of his son and could discuss "the intricacies of soil microbiology with the fluidity of a scientist." Julius told the reporter, "To think when he was a little shaver I had to go out searching for him every time a steam shovel came within two miles of our home. I was almost sure he would turn out to be a mechanic."

Albert's "young looking" mother, Rachel, "delicately hinted at items her son was too modest to mention." His older sister, Elaine, aged sixteen and now a nurse, was "trim and nice-looking," and his youngest sister, thirteen-year-old Sheila, was "determined to become her brother's assistant when she 'grows up.' "

The modest Albert, whose "eyes brighten with enthusiasm and hope" at the mention of streptomycin, had done his work "under Dr. Selman A. Waksman at the New Jersey Agricultural Experiment Station." Waksman was the director of the lab, the assigner of the task; the article did not mention him in any other role. If Waksman heard about the article, he didn't mention it.

A month later, *Collier's*, another popular magazine, ran a big feature on streptomycin, titled "Magic Germ Killer." "Streptomycin, made from a common earth mold, is the newest, safest miracle drug, effective against diseases which penicillin won't touch," the article began. There was a large photo of Waksman in his white coat at the lab bench with "his assistant Dr. Schatz" just visible behind him. When the present hunt for new antibiotics had started, the article said, the first ones discovered had all been too toxic. "Bacteriologists were wringing their hands with frustration" when Dr. Waksman spoke up about a formidable bacteria killer he had discovered and reported on twenty-nine years ago in the course of soil experiments. He called it streptomycin. "In three months—overnight in medical circles—bacteriologists were predicting that the drug would be as great as penicillin." There is no evidence that Waksman moved to curb the writer's extravagant claims in his behalf. For the time being, Schatz held his own in the trade press, however. The November 1945 issue of the *Journal of the American Pharmaceutical Association* included a thirteen-page review of streptomycin by Selman Waksman and Albert Schatz that listed them as codiscoverers, but with Waksman's name first.

Meanwhile, streptomycin was on its way to market. By the end of the year, it had moved to number three among the top ten science advances for 1945, after the atomic bomb and the large-scale production of plutonium for use in the bomb.

IN THE FALL of 1945, Waksman launched a new project in Poultry Pathology to find an antibiotic that would destroy viral diseases found in chickens. In the "gold rush" unleashed by streptomycin, the hunt was on to find antibiotics for any of the incurable diseases, including cancer. Waksman put Schatz and his friend Doris Jones in charge of the research. Jones had re-

ceived her M.A. in July after completing the first in vivo work on streptomycin. They both knew that the chances of finding a new miracle cure for
viruses—the cause of diseases such as smallpox, chicken pox, mumps,
yellow fever, influenza, and the common cold—were even slimmer than
the .1 percent chance of Schatz's finding streptomycin. He and Jones were
even moved into a different building. With increasing dismay, Schatz would
see his image of himself as the discoverer of streptomycin gradually fading as Waksman assumed the starring role.

9 · The Parable of the Sick Chicken

IT IS DIFFICULT TO SAY EXACTLY when Selman Waksman decided to rewrite the story of the discovery of streptomycin, but several strange events occurred in the Department of Soil Microbiology in the spring of 1946. Put together, they provide evidence that Waksman indeed had a deliberate strategy to downplay Albert Schatz's role, and when confronted with this charge, he did not deny it.

Looking back many years later, Waksman would describe 1946 as the year "things began to happen." And Schatz would say it was the year he "really began to feel uneasy" about how Waksman was handling the intense publicity that streptomycin was attracting. Certainly, Waksman and Schatz set out on a collision course, which would turn the mutual admiration they had for each other, and the exhilaration of the discovery, into anguish and despair for both men.

FREE OF THE patent commitment to Merck, Rutgers set up a new trust, the Rutgers Research and Endowment Foundation (RREF), to hold patents taken out by Rutgers staff. Rights to manufacture patented goods would be leased to companies in return for a royalty of 2.5 percent. Rutgers considered that it now owned the patents on the drugs being produced by Waksman's department and wanted Waksman to assign to the foundation "all improvements and future inventions." In return, the university agreed to pay him a percentage of the royalties received. There was no question of paying the graduate "assistants" who were named as codiscoverers on the patents, Boyd Woodruff for actinomycin and streptothricin and Schatz for streptomycin. It was simply assumed that because of their rank they would accept the assignment of their patents to the new foundation, and indeed, that was the protocol of the time.

Waksman was keen to do more than merely accept money. He offered to act as the manager of the patents, taking inquiries from interested companies both in the United States and abroad. Initially, Rutgers offered Waksman 15 percent of the net royalties received, after the legal and other expenses, a deal that was in line with other university patent agreements. But Waksman drove a hard bargain. In the final draft the figure "15 percent" was crossed out and replaced with "20"—the 5 percent being agreed on for the extra burden of work. He also protested the clause that included all his future inventions, and the clause was dropped.

Before they could execute the agreement for streptomycin, however, Waksman had to persuade Schatz to give up his rights to the patent, and this proved more difficult than Waksman had anticipated. On May 3, 1946, Waksman called Schatz into his office and asked him to sign the necessary papers, but Schatz hesitated. Why were they patenting a drug so badly needed by mankind? he asked. Waksman told him that assigning the patent was routine. Others had done it—Woodruff had done it for actinomycin and streptothricin. In fact, under the old deal, Woodruff had assigned his share of those patents to Merck—for whom he now worked.

Schatz said he needed to think it over, and, according to Schatz, Waksman lost his temper: "He told me that he had had enough trouble with my conceited and rebellious attitude and that I had better sign, as he had already done." Waksman demanded that Schatz sign immediately, as the document had to be returned to the lawyer that very day and sent to Washington before a U.S. Patent Office deadline. It might even be too late, Waksman said. "He said that unless I signed at once there would be no patent," Schatz recalled.

According to Schatz, Waksman told him, "Think it over for a few minutes." Schatz understood that he was cornered, but not by the Patent Office. Also according to Schatz, Waksman told him that his name would be taken off the patent application and that he, Waksman, would use his influence to "kill job chances." Schatz was afraid that Waksman could hurt his job opportunities by giving him bad recommendations, and he was not sufficiently knowledgeable about the business arrangement with the Rutgers

Foundation, or about patents, to challenge Waksman's assertions. The stakes were too high to refuse Waksman's request. Shortly, Schatz returned to Waksman's office and they shook hands, agreeing, again according to Schatz, that neither would profit from the deal; and they signed the papers. At no point did Waksman tell Schatz that he had already made a deal with the Rutgers Foundation to receive 20 percent of the net royalties.

ABOUT THIS TIME, Waksman stopped introducing Schatz to reporters and other scientists who came to visit seeking information about the discovery of streptomycin. Schatz would learn about the meetings later from newspaper and magazine reports, or from other graduate students who were working with Waksman on the third floor. Invariably, he was portrayed as Dr. Waksman's "assistant." But neither Schatz nor his family, especially Uncle Joe, was prepared to accept this downgrade, as they saw it.

In April 1946, two science publications named Waksman as the discoverer and Schatz as the "assistant." The trade journal *Chemistry* published a first-person account titled "The Story of Antibiotics." The other journal, *International Medical Digest*, a monthly journal of medical abstracts, did not mention Schatz at all, nor did it reference the two key papers in 1944 with Schatz as the senior author.

As a dentist, Uncle Joe had access to such journals, and he alerted Schatz, who wrote a letter of complaint to the editor of the medical digest. Schatz did not want to be seen confronting Dr. Waksman in public for fear that Waksman would not give the recommendations Schatz needed for future employment, so, with Uncle Joe's consent, he signed the letter using Uncle Joe's name, Dr. J. J. Martin, as his nom de plume. The letter complained specifically about the omission of the two key papers. "Streptomycin was discovered and isolated from the mold *Actinomyces griseus* by Dr. Albert Schatz," he wrote, giving references to his two 1944 papers announcing the discovery.

The editor, not knowing who Dr. J. J. Martin was, contacted Waksman to check on the complaint, and Waksman, instead of informing the editor of Schatz's part in the discovery and the importance of the scientific papers mentioned, took the opportunity to reinforce his own role and cast doubt on Schatz's. In a four-page reply, Waksman said that the discovery had come about as a result of "numerous studies," starting in 1939; these studies were the high point of "investigations of more than 30 years duration" begun by himself in 1915, and in the work he had been "assisted by hundreds of graduate students and research workers." All the steps necessary for the isolation of antibiotics had been worked out by himself and Boyd Woodruff in 1942, and that was why it had taken two years to isolate streptothricin and only "2 or 3 months" to isolate streptomycin.

About Schatz's role in the discovery, Waksman wrote, "The mere isolation of an antagonistic organism does not signify a great achievement, since a large number of such organisms have now been isolated and found to be active." The fact that he had put Schatz's name first had no significance beyond his acknowledgment of the hours put in at the workbench. "I can assure you that I would have been more than justified to have reversed the order or even to have used a common procedure used in many other laboratories, namely to place a footnote in the paper thanking the assistants for technical service thus rendered."

However, Waksman suggested, it might "have been desirable" to have mentioned two papers. One was the original 1944 paper announcing streptomycin, with Schatz as the senior author. But in place of Schatz's second key paper, dealing with the tests on H37Rv, he substituted his own paper that he had given at the Mayo Clinic, on which he was the senior author. Those two papers, one of Schatz's and one of his, Waksman wrote, "would have taken care of any criticism of incomplete references."

Then Waksman addressed the "more important view touched upon by Dr. Martin: Who discovered streptomycin?" For the first time, he concocted a story—"a parable," he called it—about the chicken strain, D-1, that Schatz had isolated from a petri dish of *A. griseus* given to him by Doris Jones. Parables are narratives of imagined events used to illustrate a moral or spiritual lesson. They are not factual. But Waksman used this "parable" in his reply, and would use several different versions of it later, to rewrite one essential act of the discovery.

The story he told began with a farmer bringing a sick chicken to Dr. Beaudette, the poultry pathologist at the Rutgers School of Agriculture. The chicken was said to be suffering from a peculiar bronchial ailment. At that time, Doris Jones was working under Dr. Beaudette trying to find an antibiotic effective against chicken viruses. After examining the chicken, Dr. Beaudette told Jones to swab the chicken's throat and put the sample on a petri dish to see if she could find any of the "antagonist organisms that Dr. Waksman is so interested in." Jones saw several colonies of actinomycetes and took the dish to Albert Schatz, suggesting that he should test them for the antibacterial properties. Schatz found one culture was active against Gram-negative bacteria. He brought the culture to Dr. Waksman, who identified the species as *Streptomyces griseus*. Waksman then instructed another assistant, Betty Bugie, to carry out certain tests that led to the "isolation and identification of the antibiotic." Later, he called in other assistants to make further tests and "thus streptomycin came into being."

So, Waksman asked in his parable, who was responsible for the isolation of the streptomycin-producing strain? Was it Schatz, was it Jones, was it Dr. Beaudette, was it the distressed farmer, was it the sick chicken? His answer: "No doubt it was the chicken, because it was she that had picked it up from the soil and started the chain of events that led to the isolation of streptomycin."

In a P.S., Waksman wrote, "I would appreciate it if the contents of this letter are not published except with specific permission."

In the years to come, Waksman would revise his "parable," each time turning it a little more to his advantage, until it became the official version of what had happened.

In May 1946, a second story cast further doubt on Albert Schatz's character. Waksman complained to his deputy, Robert Starkey, that there had been an unauthorized visit to the laboratory where Schatz and Jones were working, and that a member of Schatz's family had carried off Schatz's crucial lab notebook of the streptomycin discovery. For the first time, Waksman put locks on the lab doors. According to a letter in Waksman's archives at Rutgers, Schatz explained that he had been having trouble with his back and was in bed that day, and his paycheck was due. The checks were left in the laboratory building, and he had asked Uncle Joe to pick up his. Apparently, a staff member had seen Uncle Joe and reported him to Waksman. Schatz promised it would not happen again. That was the beginning and end of the story, as far as Schatz was concerned. As to Waksman's accusation of a missing notebook, Schatz's notebook of the discovery was not in the lab at the time of the alleged "break-in." Waksman had sent Schatz's notebook and his own to Merck, at the company's request, because it needed them to present the streptomycin patent application. The matter was dropped—but only for the time being. Waksman would bring it up again, much later.

IN YET ANOTHER incident that spring, Waksman would admit that he was deliberately keeping Schatz away from the publicity gathering around streptomycin. Doris Jones went to see Waksman on an entirely separate matter. She wanted to complain about Schatz's domineering presence in the lab where they were searching for antibiotics against viruses. He was so alert and active that she sometimes worried he might "squelch" her own ideas, she said.

In the course of conversation she referred to Schatz's "bossiness," and Waksman, to her surprise, suddenly launched into an attack on Schatz's "immaturity." It was the reason, he said, why he had decided to keep Schatz out of the story of the discovery of streptomycin. All the publicity would "go to his head." That was why he was keeping him away from reporters. He told her this "confidentially" as a way, apparently, of helping her resolve her problem with Schatz.

Jones was stunned. In her view, no one could have failed to notice how Schatz had been sidelined. Schatz himself had complained directly to Jones about "Waksie" hogging the publicity. But here was Dr. Waksman, her wonderfully erudite and supremely confident professor whom she admired so much, apparently needing to explain his actions to her, a mere graduate student. It was most out of character. It was as though he had some "hidden guilt," she thought, as if he were trying to "justify his actions to little old Doris," who could be relied upon, despite the "confidentially" admonition, to pass the information to others.

She would never recall the conversation precisely, but neither would she ever forget the incident. In fact, she did not tell Schatz for several years. Their problem was sorted out quite amicably between the two of them.

FOLLOWING THESE INCIDENTS, and with Vivian about to graduate, Schatz was ready to leave Rutgers. Waksman found him a job, almost immediately, as a senior bacteriologist at the New York State Department of Health in Albany, and gave him a glowing recommendation, quite the opposite of the confidential assessment of Schatz he had just given Jones. He told his new employer, "He is still relatively young (about 27) but he has a *mature* [emphasis added] judgment and can plan and carry out his work. I have full confidence that he will justify himself in any position of trust which will be given to him. He has been associated with me for a period of nearly 4 years . . . He has made an important contribution to the subject of antibiotics. Although he is primarily a bacteriologist, he has had sufficient training in organic chemistry and biochemistry to be able to carry through his own chemical investigations."

Schatz accepted the job and was due to depart Rutgers after four years in the graduate program. One day, however, he was talking with Waksman in his office and took the opportunity to complain, "as diplomatically as possible," that Waksman had been getting all the credit in articles about streptomycin. Schatz recalled that Waksman "became incensed and started shouting at me that he would never tolerate another Joffe"—a reference to Waksman's earlier spat with Jacob Joffe over the order of names on a scientific paper. "He then had his secretary type a letter from me to him dated May 21, 1946, telling him that I had worked under his supervision etc [and] he insisted that I sign [the letter]. I signed because I needed letters of recommendation from him when I applied for jobs."

On the morning of May 21, Waksman discussed with Russell Watson, the Rutgers Foundation lawyer, what Mr. Watson described as the "Schatz claim." Although the content of this discussion was not recorded, Watson apparently advised Waksman to get Schatz to clarify his assistant status. The result was a letter typed by Waksman's secretary, thanking him for his guidance and advice and narrowly defining his work in the discovery of streptomycin as that of an "assistant" and no more.

In the letter, Schatz expressed his "appreciation for all that you have done for me both in my undergraduate, graduate and post-doctorate work.

"I feel particularly proud to have been associated with your group in the work on antibiotics, a subject which has raised the status of microbiology to a science second to none.

"In assisting you with the isolation of the streptomycin producing organism and in the isolation of streptomycin itself, I feel that I have rendered my own contribution, no matter how small that may appear, to building and developing the science of antibiotics. I hope that the work on streptomycin, carried out under your guidance and continuous active participation, and in which in addition to myself also Miss Elizabeth Bugie and Miss H. Christine Reilly, have contributed to the best of their ability, will stand as a symbol to cooperative work under your wise and able leadership."

Schatz signed the letter, never imagining that one day it would be used against him as evidence that whatever experiments he had carried out in the basement laboratory had not really been his experiments, but had always been under the direction of Dr. Waksman.

10 · Mold in Their Pockets

NEWS OF STREPTOMYCIN'S POWER GAVE WAR-WEARY Europeans hope that they might be able to counter the sudden rise of infectious diseases, including tuberculosis. A number of factors had contributed to the increase. During the war, the TB hospitals and sanatoriums had been cleared to make room for air raid casualties, and the disease had spread more easily in overcrowded homes. Evacuations of cities to the country had introduced a whole new generation to unpasteurized milk. Workers who were desperate for jobs concealed their infections to keep working and spent longer hours in factories, reducing their levels of resistance.

All over the world, doctors were scrambling to find antibiotics to fight infectious diseases caused by Gram-negative bacteria, including urinary tract infections resistant to the sulfa drugs, various types of pneumonia, typhoid, *Salmonella* infections (also known as paratyphoid), acute brucellosis, and tularemia. Patients suffered and died for the lack of a drug like streptomycin. America's closest ally, Britain, was desperate for supplies of it.

As the world war had played a key role in the discovery of streptomycin, so now the cold war was playing a role in its development. The U.S. and British governments would continue to seek antidotes to a possible biological weapons attack, but this time from the Soviet Union. The British knew firsthand of the potential of streptomycin through secret trials at the chemical and biological warfare establishment at Porton Down. Those had shown that streptomycin was effective against the plague, endemic in some of the British colonies, and especially effective against tuberculosis. But for the moment, the U.S. government was keeping tight control over the distribution of streptomycin, as it had with penicillin, which was still available only in Allied army hospitals. In Europe, a black market of vials of penicillin flourished, including many fakes. The Russians, who went through the war with only their version of René Dubos's gramicidin, had started to produce a crude penicillin from a mold lifted from the damp wall of a Moscow air raid shelter.

As the world learned more about streptomycin, including the sweeping claims that resulted from interviews with Dr. Waksman in the popular press, American and British authorities raced to separate out the truth of what streptomycin could do before it spread into the market. For responsible scientists, the drug was still in the testing phase, but that caution did not stop desperate efforts to find supplies. It might be easier, one British official suggested, if "our people" visiting Washington were asked to bring home tiny gram amounts "in their pockets." The British ambassador in Washington, Lord Halifax, managed to find a private supply through an American doctor "well-known to this embassy," but he warned Prime Minister Clement Attlee's new Labor government that American doctors were skeptical about some of the more extreme claims. Halifax cabled London that American physicians were concerned that there had been "a great deal of erroneous information about the efficacy of this drug with the unfortunate consequence that many doctors in England who have not had the opportunity to get first-hand knowledge . . . still appear to be under the impression that the drug has far wider uses than has actually been proved to be the case in recent tests."

Streptomycin was unavailable even to some officials in high office with the right connections. Alexander Fleming and Howard Florey, who with Ernst Chain had won the Nobel Prize in Medicine for the discovery and development of penicillin in 1945, had difficulty finding supplies. American doctors and officials became increasingly troubled by the desperate calls from across the Atlantic.

There was another emerging problem: Streptomycin had uncharted side effects. In addition to temporary dizziness after prolonged administration, there were some cases of kidney troubles, and it was not clear whether these effects were due to impurities or to the drug itself.

William Feldman and Corwin Hinshaw continued to urge caution in their reports. Although it was the "most promising drug" of natural or synthetic origin against TB, they did not yet know whether a patient could

*Selman Waksman (far left) in front of one of the fifteen-thousand-
gallon fermenting tanks for mass production of streptomycin at
Merck's Elkton, Virginia, plant in December 1945.
(The Merck Archives, 2011)*

be completely cured. Moreover, it was expensive. It was generally agreed
that the minimum dose was four grams a day given in six separate doses
and that the treatment would probably have to continue for six months,
for a total cost of nine hundred dollars, a prohibitive amount for too many
patients.

The London *Times* medical correspondent was also cautious: "Great
things are hoped for with streptomycin, but early optimism must be
guarded. In any case, its use entails considerable discomfort for the patient, with six or so injections each day for possibly weeks ... Streptomycin,
which is similar to penicillin, the product of a mold and equally harmless to
human beings ... It is especially with regard to tuberculosis that experiments now made possible in this country will be watched with interest."

The British government controlled what little was available, but the controls didn't always work, of course. Some black-market supplies got through.
And the British press and the BBC featured heartrending appeals for children in desperate need of the drug. Government officials were embarrassed; the new miracle drug was in use in America, Britain's closest ally,
and none was available in Britain. "Is there really any justification for
these BBC SOS messages?" growled one government official. "... These

imply that there is no system by which the available supply can be distributed to cases of greatest need." The government feared the rise of a wider black market for streptomycin, just like the one for penicillin. In America, there were reports that such a market already existed.

The public reaction might have been more extreme if it had been generally known that at Porton Down supplies had been found to test for antidotes to biological weapons that so far had not been used, even in the last brutal months of the world war.

Merck had supplied Porton with two batches of streptomycin for its own tests on the plague, which particularly interested the British. The Porton scientists used two strains of the plague from Africa and found that streptomycin was "highly efficient and non-toxic." In a top secret report in 1945, government scientists at Porton concluded that streptomycin had great promise in combating half of the possible biological weapons then being considered. Its activity against TB was "particularly noteworthy." Each of the potential weapons was letter coded, and in total they are a measure of how far the biological weapons program had advanced on each side of the Atlantic. Decoded, with results, they included N (anthrax), "apparently highly effective"; US (porcine brucellosis), "good protection in chick embryos"; and UL (tularemia), "strikingly successful." But streptomycin had "no protective effective" against LA (glanders, which occurs primarily in horses, but has a high death rate in humans) or SI (psittacosis, or parrot disease, spread in bird feces and producing severe pneumonia-type symptoms in humans).

Of the sixteen copies made of the report, three went to Fort Detrick, the U.S. biological weapons research center, and one went to George Merck, still head of the U.S. Biological Warfare Committee. The British public was fed a different story.

The British answer to the shortage of supplies was to emphasize streptomycin's problems. A government statement to the *Times* warned that "in the very small number of patients with tubercular meningitis whose life has been prolonged by the treatment there has nearly always been permanent serious mental derangement, blindness and deafness. Steps are being taken to speed up production in this country, but not enough is known about this drug at present to justify the Government making it more freely available."

Selman Waksman had been an extraordinarily effective salesman in the U.S. media for streptomycin's therapeutic powers—and had also been nominated (unsuccessfully) in 1946 by an American physician for the Nobel Prize in Physiology or Medicine. If Waksman knew about his nomination he didn't mention it, but he was furious at the British for their public disparagement of his wonder drug. He cabled one of his contacts in the British drug industry, Sir Jack Drummond, research director of the drug company Boots, and asked him for background to the *Times* article. The situation was a "rather complicated one," Drummond replied. First, it was natural for physicians to be impatient because the drug was unavailable. Britain was behind in producing it, and there was a "real danger" that "happy-go-lucky" trials would be made in a "hit-and-miss" fashion with "little sense of scientific control or accuracy." Any supplies had been allocated by the government for official clinical trials. In the circumstances, the government's position was "a fair one, even if it erred on the side of caution."

No COUNTRY NEEDED antibiotics of all types more desperately in the postwar world than Japan. Remarkably, the Japanese had managed to smuggle scientific papers describing penicillin production through Germany, and by 1944, when U.S. marines landed in the Marshall Islands, the Japanese had started to plan their own production. But only tiny amounts were available, and they also needed streptomycin. Five hundred TB patients were dying every day, but it would be some years before Waksman and Rutgers could agree on the terms of production with Japanese companies. In the meantime, the United States began testing its new atomic bombs in the same Marshall Islands, an exercise that surprisingly led to the discovery of yet another antibiotic, just like Albert Schatz's but from a different actinomycete. This time, however, the discovery was not exactly welcomed by Waksman.

On July 1, 1946, at eight o'clock in the morning, the U.S. Air Force dropped an atomic bomb on the Bikini Atoll, in the Pacific. It was the fourth atomic detonation, after the test in the New Mexico desert at Alamogordo and the bombing of Hiroshima and Nagasaki.

Watching the test from the deck of a hospital ship, ten miles upwind, was a lanky, fair-haired Ph.D. student from Selman Waksman's laboratory named Donald Johnstone. Like Schatz, he had been drafted into the army in 1942, but unlike Schatz, he had been sent to the European front. He spent the last months of the war in a field hospital in Germany.

On his return to Rutgers, he volunteered to be a member of a team of biologists monitoring what happened to the flora and fauna of the land and waters of Bikini when the bomb exploded. The scientific team was a small part of the forty-two-thousand-strong force gathered to monitor the tests. The U.S. Navy had assembled an unmanned fleet of more than ninety vessels, including three captured German and Japanese ships, to see the effects of the blast on them.

"We were on deck around eight in the morning with our eye masks distributed by the Navy," Johnstone recalled. "They had such dark lenses you couldn't see anything. Instructions came over the loudspeaker—to put on our masks and put our arms up across our faces. At the moment of the explosion, we were told to turn away, and when we were allowed to turn back we saw the mushroom cloud."

The bomb was between fifteen hundred and two thousand feet off target and had only what the navy called a "transient effect" on the target flotilla, but it still sank five ships. About ten days later, Johnstone and the other biologists, equipped with their own Geiger counters to measure radiation, were allowed onto the island to take samples. The Pentagon was interested in what happened to the bacteria in the waters stirred up by the bomb blast, but Johnstone was more interested in the actinomycetes in the coral sands. These sands had a high pH, which Johnstone knew was ideal for actinomycetes.

In the days prior to the test, Johnstone had collected sandy soil samples and tested them in his ship's laboratory against known pathogens he had brought with him from Rutgers. He found several promising zones of antagonism, and when the Bikini bomb tests were over, he took the microbes home and grew them in petri dishes in the basement lab at the Department of Microbiology—the same lab in which Schatz had found streptomycin in 1943.

Within a few weeks, he found a culture that produced an antibiotic with almost exactly the same powers to destroy harmful microbes as streptomycin, including nonpathogenic strains of *Mycobacterium*—the same ones, from the Rutgers culture collection, that Schatz had first used against his strains 18-16 and D-1.

Three years after Schatz's discovery, Johnstone had found another actinomycete, an entirely different species, that apparently produced streptomycin. In fact, in all of his two or three hundred cultures from Bikini, Johnstone never saw a single *A. griseus*. And his discovery was nothing to do with the atom bomb radiating microbes on the beach and producing mutants that then produced streptomycin. He had collected his cultures *before* the blast.

Waksman told him to check his experiments, just as he had told Schatz to check his on *A. griseus*, and then write them up for publication. They agreed on a name, *Actinomyces bikiniensis*, and they called the new antibiotic streptomycin II. On May 15, 1947, Johnstone announced the discovery at a meeting of the Society of American Bacteriologists in Philadelphia. SCIENTIST TELLS OF NEW DRUG: STREPTOMYCIN NO. 2 CAME FROM BIKINI, reported the *Philadelphia Evening Bulletin*. Like its predecessor, the newspaper said, it worked against the tuberculosis germ.

But Johnstone added that his streptomycin II was slightly better than streptomycin I. Two-tenths of a unit of streptomycin I were needed to accomplish the same effect against the TB germ as was produced by one-tenth of a unit of streptomycin II.

After the meeting, he suddenly found himself surrounded by drug company representatives wanting him to go work for them, but "I told them no, I wanted to stay in basic research at a university," he later recounted.

When he got back to Rutgers, Dr. Waksman "was beside himself," Johnstone recalled. Waksman told him that the newspaper stories were "embarrassing." The composition of the new drug that Johnstone had discovered was not yet known; it looked like streptomycin, but no one yet knew whether it was really the same thing, as the reports had said.

"You better get out there and give the newspapers an account so you won't embarrass the rest of the world," Waksman said. Johnstone followed his professor's order: "I gave a report blasting the [original] report that it was something new and better."

What Johnstone did not know at the time was that Waksman was in the middle of applying for a "product" patent for streptomycin. Waksman was asking the patent examiner to grant rights for the substance streptomycin, as produced by *A. griseus*. He apparently did not want to be distracted by another, similar discovery.

11 · Dr. Schatz Goes to Albany

THE SAME MONTH AS THE BIKINI test, July 1946, Vivian graduated from the Women's College. Uncle Joe bought Schatz a secondhand car, and the young couple made their way to Albany, in upstate New York, where Schatz would start his new career as a civil servant employed by the state's Department of Health. Robert Clothier, the president of Rutgers, gave Schatz a farewell letter expressing his "sense of regret" that Schatz was leaving. "It has been a source of great satisfaction, both official and personal, to have had you associated with us and I hope that continued success and happiness will be yours." Schatz was pleased to have the letter even though he realized it was mostly a formality. A more genuine letter came from a member of the staff of Plant Pathology, who recalled that Waksman, "in the presence of several others," had said, "Schatz was the most brilliant student I ever had." It was said "in all sincerity."

But there was no farewell letter from Waksman himself, and Schatz had not expected one. Uncle Joe had been keeping his vigil on the media. *Time* magazine had run a story on "Streptomycin Wonders" with the news that the drug was now being distributed to sixteen hundred U.S. hospitals. It was "a triumph for the drug's discovery in 1944 by Rutgers's microbiologist Selman A. Waksman." Schatz wasn't mentioned. The *New Jersey Journal of Pharmacy* had run a story on the discovery after interviewing Waksman. The story described the discovery as a "product of extensive studies in soil microbiology by Dr. Waksman." Using his Uncle Joe nom de plume, Schatz complained, and the journal obliged with a correction. While the credit it had given Waksman was accurate, it "does not mean

The staff of the department of microbiology on the steps of the administration building after Albert quit Rutgers. Selman Waksman is in the front row wearing a bow tie, with Dr. Starkey (left) and Dr. Geiger (right). In the back row is Donald Johnstone (left), and in the next row, fourth from the left, is Doris Jones (ca. 1947). (Special Collections and Archives, Rutgers University Libraries)

that sole credibility for the discovery . . . belongs to him. Dr. Waksman would be the last to make such a claim . . . Today, practically no major discovery in science is the product of one man's work." If Waksman saw the correction, and Schatz was sure that he did, as he read everything, then he made no comment.

SCHATZ HOPED THAT the storm over the patent had blown over. Waksman had not mentioned it again, and Schatz was keen to forget it and begin his new life with Vivian away from Rutgers. He started a regular correspondence with Waksman, writing at least once a month, telling him how he was getting on and sending "regards to the group." He even asked Waksman for a photograph that he could put on his desk, and Waksman sent him one.

These letters were always friendly and personal. Schatz would tell Waksman if he felt he was not doing so well. He found he was not good at handling pressure from above. "I don't mind work and I have not been loafing here by any means. But it 'feels' different when I drive myself. But I suppose this will always be with me, so I'd better get used to it."

Waksman responded with concern and advice. He suggested that Schatz belonged in a "pure research organization"—Caltech, in Pasadena, for example. He should have "no difficulty" in getting a fellowship from the U.S. Public Health Service, or Merck. Or Schatz might like to take a year off in Europe, or even Russia, if the situation improved sufficiently to welcome foreign visitors.

ALBANY WAS CERTAINLY not the expected career move for a brilliant student who had played a key role in the discovery of the world's most-talked-about new miracle cure. The city itself was dull. The capital of New York State lived in the long shadow of New York City, 150 miles to the south, and was the seat of a corrupt state legislature.

Even so, the Health Department's Division of Laboratories and Research, where Schatz worked, had an international reputation. Known as the "Division," it was founded in 1880 as a combination of a scientific laboratory doing pure and applied research, an educational institution, and an operating public health service. It was respected by scientists worldwide. The Division's director, Gilbert Dalldorf, who had joined in 1945, was concerned that the successes of penicillin and streptomycin against bacteria would make public health researchers complacent about diseases caused by fungi and especially viruses, for which there was no cure.

Dalldorf had led a "determined effort" to establish virus studies at the Division at the beginning of 1946, and Schatz joined the virus team. But funds for the Department of Health were limited. Schatz found the labs poorly equipped, lacking enough egg incubators and mice cages to do adequate experiments. Plans for a new virus building kept being postponed because real estate prices were high after the end of the war. Another reason was that Governor Thomas Dewey was planning a run for the presidency against Harry Truman, and with his eye on the 1948 election, he was using the state's coffers for political favors, not the civil service, which faced pay and staff cuts.

Schatz kept Waksman informed of these problems, and Waksman offered fatherly advice and urged him not to be discouraged. He recalled the shortages at Rutgers after World War One: "We almost had to make our own test tubes and petri dishes." Any beginning was difficult, he counseled. "One should figure about a year before the laboratory is well-organized."

Overall, Waksman must have been relieved that Schatz had left Rutgers—he could now make his own way with streptomycin—and Schatz had readily accepted the Albany job because he was keen to get out from under Waksman's wing. He was also now married and in need of the money. Pure research, unattached to business or the government, was what he would have preferred to be doing, but in 1946 such work was barely possible in America without a fellowship, and while Waksman had mentioned the idea of a fellowship, he had not offered to help him find one.

The ever-restless Schatz turned to other things. "I'm determined to get a general cultural education," he wrote Doris Jones. "This business of science, science, straight science, technical stuff and more technical stuff seems to make scientists among the dullest people in the world. There are different people, different ways of life, different philosophies, there are histories and arts. I shall now begin to learn some of this for man is emotional as well as intellectual . . . To develop one of these aspects without the other is to grow lopsided." He told Waksman he was determined to become fluent in Russian. He had translated one Russian paper, only a page long, but it had taken him a whole day.

Vivian found a job as a bacteriologist in a local hospital, but when Albert pointed out how unsanitary the place was, she quit. She found a different interest. Paul Robeson had been invited by Albany's Israel African Methodist Episcopal Church to sing at a local school, but the city had refused to issue a permit because the House Un-American Activities Committee had linked Robeson to communists. Robeson was a Rutgers graduate, and Vivian joined the Let Paul Robeson Sing Committee, running flyers around town. A local judge ruled that Robeson could sing, provided he did not discuss politics. The FBI was following Vivian and had opened a file on her, though she would not know this until seven years later. In 1952, when she and Albert were living in Philadelphia, the FBI knocked on their door and asked Vivian about her colleagues on the protest committee. She told them she could not remember anything.

BY THE BEGINNING of 1947, Schatz was thinking about a new job. He felt pressure to produce something on viruses, but did not have the means to do it. He had heard through the "grapevine" that he had not made the progress his supervisor had hoped for, he told Waksman, and he was spending most of his time isolating organisms antagonistic to pathogenic fungi. He was only "nibbling" at the bigger virus problem. "To be honest with you, I am not very much interested in these problems, but I am doing them because I must do something." The equipment situation continued to interrupt his work. Even the supply of eggs was poor. "This morning was heartbreaking—12 good eggs alive out of 200!!" He was working hard. "Last weekend, I was in the lab for about 16 hours and I frequently stay until midnight, and then up again at 4 A.M."

No doubt, Waksman had the contacts at Merck to arrange a fellowship at the company or another academic post, but he never actually came up with anything. Perhaps the problem was what Schatz might learn from Merck about the work of its researchers in his antibiotics project, or, worse from Waksman's position, what Schatz might tell Merck about the discovery of streptomycin. At any rate, he instead steered Schatz to a job at the Sloan-Kettering Institute, in New York City, where Waksman was a director. Schatz would now be looking for antibiotics that would destroy cancer cells. The task had a dreamlike quality about it, seemingly impossible, but the pay was good: five thousand dollars a year—almost twice what he was getting in Albany. And Sloan-Kettering provided housing in Manhattan.

Before leaving Albany, Schatz was invited, as a result of his employment in Albany, by the New York Association of Public Health to give a talk on the history of streptomycin. It was his first chance to publicly put forward his view of the discovery, and he used the opportunity to construct a markedly different version from Waksman's claims of a long-term, systematic search among tens of thousands of cultures. Instead, Schatz modestly acknowledged how fortunate he had been in finding a new antibiotic so quickly. And he was very careful to avoid using the first person, so that his five-page report always referred to the laboratory effort, not just his.

"Nothing in science begins *de novo*," and this was "particularly true in the recently developed field of antibiotics," Schatz began. Even the term "antibiosis," he noted pointedly, was introduced in 1889 by Paul Vuillemin, to designate an antagonistic state of living organisms. "Consequently, the discovery of streptomycin involved nothing new in principle." Schatz had selected his cultures almost entirely at random—"simply because they appealed to the speaker." This was a direct attack on Waksman's claim that Schatz would never have found streptomycin if he had not followed methods already in place for the earlier antibiotics. Schatz insisted that his own work had illustrated the "fortuitous nature" of antibiotic searches. "It has been frequently pointed out that penicillin was discovered as the result of an accidental observation, whereas other antibiotics, such as streptomycin, resulted from organized research with specific objectives. Such an attempted differentiation appears more philosophical than practical," he wrote. "The differentiation which is of importance is that penicillin, discovered in 1929, had to wait almost ten years for the interest of the medical world, whereas in 1943 an interest was awaiting streptomycin— Merck."

At the time, Waksman saw only an abstract of the paper and again had no comment. But Schatz, despite his civility and deference in his letters to Waksman, was increasingly aware that something was deeply wrong in their relationship. The row over the patent had left him with more than a bitter taste. He could not avoid the feeling that he had been tricked in some way that he did not know or understand. Uncle Joe encouraged him to keep up his anonymous attacks in the media on Waksman's claim to be the discoverer, but Schatz wanted more than that: He wanted to find out why Waksman had flared up when he had hesitated to sign the patent documents. The five-page paper he had just written, he realized, had been a cathartic experience. He had told the truth about the discovery as he knew it. What was Waksman concealing? Schatz did not relish a confrontation, but he felt that his curiosity might soon get the better of him.

12 · The Five-Hundred-Dollar Check

![ornament: microscope] **BY 1947, THE AGE OF ANTIBIOTICS,** like the big band era, was in full swing. Sales of penicillin and streptomycin were booming, but with penicillin production still four times that of streptomycin. The world's largest penicillin maker, Chas. Pfizer & Co. of Brooklyn, New York, had posted a ten-million-dollar profit on 1946 sales of about forty-three million dollars. At the end of 1946, the U.S. government lifted controls on streptomycin distribution. Under government control three quarters of streptomycin production had gone to the armed forces, the Public Health Service, and the Veterans Administration. Clinical tests had produced positive results in both miliary tuberculosis, where the germs ride the bloodstream and lodge in body organs, and tubercular meningitis, where the germs attack the brain and spinal cord. The other quarter had gone to the National Research Council, which had carried out separate trials on one thousand patients with bacterial infections. The council's official report concluded that streptomycin could definitely arrest tuberculosis, but it would not necessarily eradicate the disease. It was an almost sure cure for tularemia, urinary tract (kidney and bladder) infections, bloodstream infections, and most cases of peritonitis and lung infections.

The popular media picked up the report. A February 1947 issue of *Life* magazine carried a dramatic illustration of a petri dish containing streaks of bacteria which cause TB, typhoid, pneumonia, tularemia, urinary infections, and staphylococcus. When treated with streptomycin "discovered by Dr. Selman A. Waksman of Rutgers University," all except staphylococcus infections were cured. "Fortunately," the article noted, staphylococcus infections were cured by penicillin. The media reports were not all positive, however. Streptomycin was expensive, at sixteen dollars a gram, with an average treatment of six to eight weeks. The issue of TB germs building up resistance was also a worry. IS STREPTOMYCIN THE ATOM BOMB IN TB WAR? asked the *New York World-Telegram*. The article said scientists were studying why streptomycin had proved to be a dud in some tuberculosis cases because of the germs' buildup of resistance to the drug. In other cases, doctors were still unsure of the right dose. "So many milligrams cured a two-pound guinea pig [but] for a 100-pound man 50 times as much might be the right dose." In addition, streptomycin was also more difficult to produce than penicillin. The yield per quantity of fluid was much smaller, and about ten times as much streptomycin by weight was required to treat a patient.

Even so, eleven drug companies, led by Merck and including Pfizer, Squibb, and Eli Lilly, had been licensed by the Rutgers Foundation, a process overseen by Waksman to produce streptomycin. The companies were investing millions in production plants. Merck was the first and by far the largest streptomycin producer from its Elkton, Virginia, plant. At the end of 1947, Merck sent the foundation its first royalty check—2.5 percent of sales. The check was for $344,907, covering all sales, including to U.S. government agencies, from June 1945. Waksman's 20 percent, minus the foundation's expenses, came to $53,192—five times his salary.

At the end of January, Waksman asked Schatz to meet him for coffee and cake at a deli in Manhattan. After some chat about Schatz's new job at Sloan-Kettering, Waksman pulled out his wallet, took out a check for five hundred dollars, and pushed it across the table to Schatz. It was drawn on Waksman's personal account at the National Bank of New Jersey.

Schatz was dumbfounded and refused to accept the check. He didn't need the money, he told Waksman. He was making five thousand dollars a year at the Sloan-Kettering Institute, which was more than enough to cover his expenses, even living in Manhattan. But Waksman insisted, keeping the check on the table and explaining that the Rutgers Foundation had given him some money and he wanted to share it with Schatz. He regretted that the foundation had not seen fit to reward Schatz directly for his work on streptomycin.

Embarrassed, Schatz struggled for a way out. If Waksman meant that a share of the streptomycin royalties should be his, then he would rather have it in the form of a fellowship from the foundation, not from Waksman personally. Without giving a reason, Waksman said that it was impossible to recommend him for a fellowship, and he insisted that Schatz accept the check. To avoid further offense, Schatz took it. They parted, and Schatz cashed the check a week later.

Waksman had caught Schatz at an especially vulnerable moment. Things were not working out at Sloan-Kettering. After three months, Schatz had still been unable to move into his laboratory because it hadn't been ready, and he had busied himself by taking a biochemistry course at New York University and a scientific Russian course at Columbia. In a letter, he told Waksman that he wanted to like Sloan-Kettering "and profit considerably in many ways," including from the facilities not available in Albany. But somehow it hadn't happened yet.

And instead of getting better, things seemed to get worse. He was told he could work on "anything I want to." He wanted "to work desperately, especially after the year in Albany," he had the equipment and an assistant, but somehow he couldn't get started.

To one of his letters to Waksman, he attached an "Epilogue," which he titled "Peregrinations of a Young Man's Fancy."

Since taking my degree in 1945, I feel like Tennyson's Galahad, I have been chasing "wandering fires." A little phage work [a phage is a virus that infects bacteria], an abortive approach to viruses, and now a fading touch of cancer do not add up to much. Doing nothing does not bother me half so much as learning nothing.

My having left A. *griseus* and streptomycin in 1945 was a blunder of the worst kind . . . Such things, however, must unfortunately be learned the hard way . . . The fact that I am not connected with anything now does not concern me very much . . . Getting into the medical field was another [blunder] . . . I have not the slightest desire for fame, glory, popular acclamation, or a lot of money. I want to do the work I like and feel good about it "inside of me." I think you know what I mean better than my feeble attempt to get it down into words.

Schatz said he wanted to go back to bacteria research and asked Waksman to help him get a place with Dr. C. B. van Niel at the Hopkins Marine Station, attached to Stanford University and located in Pacific

Grove, California, near Monterey. Van Niel was a Dutch-born bacteriologist who had made his reputation during the 1930s on the metabolism of many groups of microbes but especially photosynthetic bacteria. By the early 1940s, van Niel's summer course attracted many more applicants than could be accepted. He was known for his strong preference for the heretical over the conventional and for being an electrifying teacher; his lectures lasted for several hours, holding his students spellbound. Pacific Grove had become the mecca of American microbiology. In teaching and in science, van Niel's method was the opposite of Waksman's textbook approach—often Waksman's own textbook, which he would read aloud in class, with no comment before or after. Van Niel, by contrast, provided advice, inspiration, guidance, and criticism, and he was "not one to use his students or co-workers as helping hands in achieving his own ends."

Van Niel's approach was a perfect fit for Schatz. "That is what I should have done originally after receiving my degree," he wrote to Waksman. "Since then my pocket and stomach have been full, but my head and heart have become empty. The happiest years of my life were spent when I was an undergraduate student with you. I felt I was doing something and I loved the work. Albany isn't worth mentioning!! Here [at Sloan-Kettering] I have done nothing since September."

He wanted to apply for a fellowship to study under van Niel. If he couldn't get one, he wanted to go anyway. He and Vivian had saved "close to $1,000." He was sure they could borrow another thousand or so if necessary. "I'd rather work for nothing in a Department such as van Niel's than at 10 or even 100 times my salary in a place such as this (unless I have erred considerably in the conclusions I have drawn)."

Waksman replied immediately, expressing surprise at Schatz's reactions to Sloan-Kettering but adding that he "could well understand" his feelings and that if Schatz wanted to work with van Niel, he would do "everything I can to help you . . . I hope that he will find room for you in his laboratory." As to the fellowship, he would "be willing to make a special recommendation to the Rutgers Foundation that an exception be made in your favor in order to grant you a post-doctorate fellowship" to work off campus. "Your financial situation will thus be taken care of." The "important thing" was to gain admission to van Niel's lab.

MEANWHILE, THE RUTGERS PR Department had been busy burnishing Waksman's image. It was staging a radio drama of the discovery of streptomycin, starring Hollywood actors, to be aired from the Rutgers campus. The press release referred to "Dr. Selman A. Waksman, Rutgers University's world famed scientist, whose discovery of the wonder drug, streptomycin, is celebrated in the drama, 'Winner Takes Life,' a *Cavalcade of America* broadcast, sponsored by the Du Pont Company."

Waksman was to attend the first airing, on April 14 and 15, and meet the Hungarian-born film star Paul Lukas, who would play him. Lukas had won an Academy Award for his role in the 1943 movie *Watch on the Rhine*, in which he had played an American working against the Nazis. He was also a charter member of the Motion Picture Alliance for the Preservation of American Ideals, a group that lobbied against communist influence in Hollywood.

More than twelve thousand people were expected to attend the four performances at the Rutgers University gymnasium. Free tickets were being offered to local companies. Employees at five DuPont plants in New Jersey, together with their families and friends, as well as students and faculty of Rutgers, received tickets. Albert and Vivian Schatz were not invited—even though "Al Schatz" appeared as a character in the show, and he was living in New York City, an hour's train journey away.

THE PLAY OPENS toward the end of World War Two, with an army general and Waksman discussing the discovery of streptothricin. It looked "very promising," Waksman tells the general, but unfortunately it turned out to be toxic. In the next scene, Waksman is talking to "one of his assistants, Al Schatz," in their laboratory, discussing how to find new antibiotics among the actinomycetes. Together they look at several cultures of *Streptomyces*, some of which produce antibiotics that "pack a wallop." In a separate scene, at a military hospital, a young soldier is dying from a bacterial infection and desperate for a new medicine. Back in the laboratory, Waksman and Schatz continue their experiments and find one strain that produces an antibiotic effective against TB. Waksman warns Schatz this is only the beginning, more tests must be carried out. He calls in another "assistant," Miss Bugie, to work on the problem. Soon they have enough of the new drug for trials on mice. It works. Mice with *Salmonella* intestinal poisoning recover. They look "frisky as a squirrel," and the untreated mice die. Waksman names the new drug streptomycin. In the final scene, the sick soldier is given streptomycin and also makes a miraculous recovery.

IN THE FALL of 1948, Schatz joined van Niel's lab, but without the Rutgers fellowship that Waksman had indicated he might be able to arrange. In fact, there is no evidence that Waksman ever asked for one. Schatz relied instead on the GI Bill for ex-servicemen, which gave him about one hundred dollars a month. Waksman offered additional support from "our own funds," if Schatz needed it. Waksman had just received a second personal royalty check for $89,617, making a total of $142,809—more than ten times his salary.

At the end of August, he sent Schatz another check for $500. Again, Schatz thanked him profusely, but he said that he could accept it only as a loan. With their savings and the GI Bill assistance, "we ought to be able to skimp along." He appreciated Waksman's kind offer for the remaining two trimesters, but could not bring himself "to impose further upon your generosity unless my own funds become exhausted."

Schatz could not have been happier with his move to the West Coast, where some of the brighter stars of American microbiology worked with Dr. van Niel. He had finally settled on his project: *Hydrobacteria*, a weird group of microbes that feed on hydrogen instead of carbon or nitrogen, like most of their cousins. Schatz's task was to isolate these hydrogen eaters and find out how to use them to create water, or moisture, in enclosed environments. He was also upgrading his skills in classes on calculus and physical chemistry and giving himself courses in thermodynamics and theoretical organic chemistry.

As he had no fellowship, the GI Bill check was his only steady source of income. Vivian was working about fifteen hours a week keeping the laboratory glassware in order, which brought in a little extra. Together they were "in good financial condition" and foresaw "no difficulties for the remainder of the year." Leaving Sloan-Kettering had been the right thing to do. In spirit, he was as "rich as Croesus."

Going after jobs that paid a decent salary had been a mistake, Schatz wrote Waksman. He blamed himself for not listening to his former professor. He recalled that Waksman had once advised him to take a course with van Niel. "The decision was entirely my own and I have only myself to blame."

Now he and Vivian were "both fine, physically, financially, mentally and in all other ways." They were "collaborating on a most important experiment"—their first child, due the following year. Schatz spent all his time in the lab, as he liked to do, obsessed with his new project to the exclusion of all others, including any thought of streptomycin.

When Waksman sent him a copy of the recently published collected scientific papers on the drug, Schatz replied that he did not have the time, or need, to follow this field as closely as before. "I simply had no idea that so very much had already been published on streptomycin," he wrote. He was working so hard that he had exhausted the supply of flasks in the lab and "consequently [had] decreed this Sunday a holiday to be enjoyed away from the laboratory." This was a first since he had arrived at Pacific Grove, a regular day off.

Schatz did get himself in the local newspaper, however. The *Monterey Peninsula Herald*'s Round and About columnist, Ritch Lovejoy, went to see him. "Last week," Lovejoy wrote, "I went over to Hopkins Marine Station to meet Dr. Albert Schatz, 28, jet-black-bearded, short, husky and largely responsible for the discovery of Streptomycin." It was at Rutgers University, Lovejoy informed his readers, that in "collaboration with Dr S.A. Waksman and Miss Elizabeth Bugie," Schatz had developed streptomycin. Although he would have liked to tell his readers more of this important discovery, Lovejoy added, the fact that Schatz was "pretty tired of hearing about it led us to drop the subject."

Schatz was now "devoting his time to microbial physiology," mainly with his *Hydrobacteria*. "A number have been isolated before, and whether the ones Schatz is watching have been described before, he doesn't know yet. He had found two—one shaped like a rod with a tail, and one without a tail. The tailed microbe travels around and the other just floated lazily in the water. They both burn up hydrogen, just like a car burns up gas and you and I burn up carbohydrates. In all cases, the result is energy."

Waksman apparently saw this article. It must have been sent to him by someone in the lab, because a copy is in his Rutgers archive. He made no comment.

———

IN THE MEANTIME, Seymour Hutner, a microbiologist at the independent research unit, from Haskins Laboratories in New York City, had alerted Schatz to a vacant teaching and research position at Brooklyn College to follow his stay at van Niel's lab. Schatz got the job—this time without Waksman's help. But Waksman was encouraging, as always. He was "delighted indeed" to learn about Schatz's progress and "particularly pleased" to learn of his Brooklyn appointment. "It certainly does credit to the members of the appointment committee of that institution who were able to recognize your merit for the position in question and to give you the proper appointment of Assistant Professor to which you are fully entitled."

13 · A Patent That Shaped the World

ON SEPTEMBER 21, 1948, U.S. PATENT No. 2,449,866, for "streptomycin and process of preparation," was granted to the inventors Selman Waksman and Albert Schatz. Of course, the "inventors" had already assigned their rights to the Rutgers Foundation, so they in fact owned no part of the new drug. And the four-page document, written in the dry, technical language of patent law, gave no hint of the division of labor. In case anyone should be in any doubt, however, as to which of the two inventors to attach greater importance, the first paragraph stated that streptomycin came from strains of the microorganism *Actinomyces griseus*, which was "first isolated from the soil and characterized by one of the present applicants, S. A. Waksman, and is described in his publication in Soil Science 8, 71, (1919)." In other words, without Waksman's earlier discovery, the present one would not have happened.

In reality, the Russian researcher Alexander Krainsky had been the first to identify the organism *A. griseus*, in 1914. And Dr. Waksman's strain of *A. griseus* was not a streptomycin producer. But Waksman had made the claim in his own behalf so often now that no one complained. This was the cold war. The patent was already paying royalties—and would continue to do so for the next decade or so, until it ran out. Waksman, and his heirs after his death, were guaranteed a 20 percent share of whatever the Rutgers Foundation made from its 2.5 percent royalty. The professor would become rich as well as famous; the Rutgers Foundation coffers would be filled in a manner that the tiny college had never dreamed of. And Schatz, Waksman's "co-inventor," would get precisely nothing. He had not even been paid the one dollar due to him for signing over the patent to the Rutgers Foundation in 1946.

Neither Waksman nor the foundation would notify him that the patent had been granted, a milestone in the history of medicine that would be hailed in 1961 by the *New York Times Magazine* as one of the "ten patents that shaped the world," along with those for moldable plastics, gasoline, the telephone, Edison's lamp, rockets, man-made fibers, powered flight, the vacuum tube, and atomic power.

Waksman had no intention, either now or later, of informing Schatz; such information in the hands of his diligent but unpredictable researcher might encourage awkward questions as to where exactly the royalties were going, not to mention the provenance of his five-hundred-dollar personal checks to Schatz.

THE STREPTOMYCIN PATENT was indeed poised to shape the future of the drug industry. There had been no patent on penicillin. In a very British manner, Alexander Fleming had never considered applying for a patent for a drug so desperately needed by humanity. Howard Florey had opposed Ernst Chain on the same grounds when Chain had suggested they should seek a patent. Even if they had applied, though, the application would have been rejected. Under British and U.S. law, patents were not awarded for products of nature, which everyone acknowledged penicillin to be. Moreover, a patent had to be applied for within a year of the discovery's public announcement. Penicillin was announced by Fleming in 1929, but by the time it was "rediscovered" and finally produced in 1940, more than a decade had passed. It was out of time.

However, the discovery of streptomycin demonstrated that Fleming's penicillin was not an isolated phenomenon; there were billions of microbes lurking in the soil, the water, and the air, in sewage plants and compost piles, and even in the gullets of chickens, and any of them might provide a wonder drug. Until now, the pharmaceutical industry had operated with a poorly stocked medicine chest of fluids, ointments, snake oil, and exotic-plant extracts. None of those—morphine, quinine, digitalis, insulin, codeine, aspirin, arsenicals, nitroglycerin, and compounds of mercury—was a cure, only a palliative. In contrast, antibiotics actually cured people of deadly infectious diseases, and the streptomycin patent showed the way. It meant companies that discovered antibiotics stood to reap handsome profits. Selman Waksman, now a scientist-entrepreneur, played a crucial role.

When Waksman first applied for patents for his antibiotics found at Rutgers, the U.S. Patent Office examiners raised the "products of nature" issue. All courts, including the Supreme Court, agreed that anything made by or found in nature—a metal, for example, or a plant, a piece of wood, or a natural dye—was in the public domain, and no inventor or discoverer could claim it as his own. However, there was no precise definition of a product of nature, and no agreed-on point at which a natural product, changed by human hands, ceased to be natural. On this key matter the courts were silent.

Waksman had some experience with this debate, from when he worked for the Takamine Laboratory from 1918 to 1920. In 1903, Takamine had won a patent case concerning adrenaline, which was a "natural" product of adrenal glands from slaughtered sheep and cattle. The company's success would ease the passage of patents granted to the future antibiotics industry and to today's biotechnology companies. The company's founder, Japanese-born Jokichi Takamine, would become known as the "pioneer of American biotechnology."

Back then, he licensed his adrenaline invention to America's first drug manufacturer, Parke, Davis & Co. of Detroit. The company started producing the drug, which was written up in the media as one of the new so-called blockbuster medicines. It helped control excessive bleeding in surgery and was also used in cardiology, obstetrics, and the treatment of asthma and other allergies. (It is still used today to relieve breathing difficulties.) As the medical uses became apparent, Takamine obtained five separate new patents, including ones in Britain and Japan.

Eight years later, another drug company, H. K. Mulford, a maker of aphrodisiacs and talcum powder and a minor enterprise compared with Parke, Davis, challenged the patent mainly on "priority" grounds. Mulford considered that it had been the first to isolate adrenaline from its natural state.

A furious legal battle ensued, with both sides engaging in several days of technical debate. The adrenaline case was heard in New York by Judge Learned Hand, a jurist with a reputation for legal craftsmanship and clarity, who was then on the U.S. District Court for the Southern District. Judge Hand relished the complexity of the case.

When Takamine had originally applied for a patent, the examiner had rejected his claim because he believed that no product patent was possible if the discovery was merely separated from its natural surroundings and remained unchanged. To get around this objection, Takamine changed the chemical nature of the extracted adrenaline, turning it from a salt, which is the form naturally found in the suprarenal gland, into a base.

The secretion from the adrenal glands in situ was clearly a product of nature, but Judge Hand steered the court toward a separate and fundamental question: whether a patent could be issued on a product of the adrenal glands that was purified for therapeutic use.

Judge Hand declared that what Takamine had created was an important distinction; the base was an original production of Takamine's. He was also the first to make the drug available for any use by removing it from the gland tissue in which it was found. Judge Hand wrote that it became "for every practical purpose a new thing commercially and therapeutically. That was good ground for a patent."

Although he had not been asked specifically about adrenaline as a product of nature, Judge Hand became well known for his ruling that purified products of nature can be patented. Waksman did not refer to his knowledge of Takamine's patent, but it must have boosted his confidence when he came to argue the patent application for his first antibiotic, actinomycin, in 1940. The Patent Office ruled immediately that the applicants (Waksman and Boyd Woodruff) had failed "to adequately distinguish" between actinomycin and products found in nature. The patent examiner wrote, "It is expected that actinomyces will generate (in natural cultural medium) antibiotics. Products of nature are 'old' and the fact of their existence cannot be claimed per se."

Waksman was ready with a counterargument. Antibiotics are only produced in artificial nutrients on the laboratory bench, he asserted. In an amendment to the actinomycin application, he argued, "It is not uncommon for organisms, when cultivated in artificial media, to produce substances which are *not* [emphasis added] produced in any detectable amounts by the organism as it occurs in nature."

Any "lingering doubt," he added, in legal language that clearly came from Merck's lawyers, "should be resolved in the applicants' favor." Backing the application, Merck lawyers argued that the Patent Office objection was "unsupported either by the cited art or by affidavit as to facts within the knowledge of the Examiner." The application was granted. And so was the one for Waksman's second antibiotic, streptothricin, after Waksman made similar arguments.

In considering the application for streptomycin, the Patent Office raised the same "product of nature" question yet again. Waksman, with two victories behind him, was more confident this time. In a much bolder assertion, he declared, "Various attempts made in our laboratory to isolate it or demonstrate its presence in the soil failed to detect it. This antibiotic is not a product of nature, but produced by particular strains of *A. griseus* only in cultures of given composition and under given conditions of culture as outlined in the patent application."

Waksman proposed his negative test—"We did not find any streptomycin in the soil"—as proof that antibiotics were not found in nature. He thereby shifted the burden of proof onto the Patent Office, which promptly surrendered. The examiner, apparently unwilling to challenge Waksman's prestige and honors in the subject, granted the patent.

BUT THE EXAMINER would not have had to look far for evidence to challenge Waksman's assertion—in fact, no further than Waksman's own book on microbe wars, first published in 1945 and reprinted in 1947.

In a passage discussing the destruction of disease bacteria in the soil by actinomycetes, Waksman wrote, "The fact that many pathogens can grow steadily in sterilized soil but do not survive in normal soil tends to add weight to the theory of the destructive effect upon pathogens of the microbiological population in normal soil." In other words, disease bacteria in natural soil seem to be destroyed by other bacteria, by fungi or actinomycetes that produce antibiotics.

The Patent Office examiner could also have found evidence challenging Waksman in the work of other researchers. The key phrase in Waksman's presentation to the Patent Office was "in our laboratory." Waksman knew perfectly well that others had reported antibiotics in nature—in the soil. Prime among them were the Russian researchers whose work he also cited in his 1945 book.

In 1937, one Russian researcher reported that "the antagonistic effects of actinomycetes were manifested not only in artificial media, but *also in*

soil [emphasis added], the interrelations here being much more complex."
The antagonistic action was more intense in light (podzol) soils and was
greatly reduced in heavy black earth, or chernozem soils—one of the factors
apparently being the high content of organic matter in the chernozem. According to the Russian research, antibiotics were produced in nature, but at
different rates, depending on the environment. To say that they were not
produced in the soil was incorrect.

Many of Waksman's Western colleagues simply disagreed with him,
but then, they were not applying for patents. In the coming years, Waksman was to hold fast to his assertion and castigate those who suggested
that antibiotics were produced in the soil, albeit in minuscule quantities.
At other times, Waksman's reply to his colleagues on this matter sometimes
seemed crafted for the moment. "At times, when Waksman was giving a
talk, he would say that antibiotics are weapons in the struggle for existence
among microorganisms. At other times he would say antibiotics only exist
when you grow microorganisms in a man-made medium," recalled one of
Waksman's graduates.

In dropping its objection to the "product of nature" rule, the Patent Office
also allowed a so-called broad patent on the product itself, streptomycin.
The patent covered the chemical compound streptomycin, however it was
made—by *A. griseus* or another microbe or synthetically in the laboratory. The Rutgers lawyers congratulated themselves. Russell Watson, the
lawyer for the Rutgers Foundation, bragged about it to one company seeking a license to make streptomycin from *A. griseus*: "We are particularly
pleased with Claim No. 13 which, as you will observe, is a product patent
expressed by the single word 'streptomycin.'" And the foundation's patent
lawyer told Watson the product claim would "prevent the importation
into this country of streptomycin manufactured abroad."

In a separate application by Merck researchers, argued by the company's lawyers, the Patent Office granted a patent for the method used by
Merck's chemists to extract and purify the drug. In 1945, Merck researchers
had claimed a patent on "complex salts of streptomycin" and the process for
preparing the drug. Merck lawyers argued that this method was patentable because the therapeutic benefits of the drug were unknown prior to
its isolation, echoing Judge Learned Hand's ruling in the Takamine case.
"Thus, for the first time," a Merck lawyer asserted, "streptomycin is available in a form which not only has valuable therapeutic properties but also

can be produced, distributed, and administered in a therapeutic way." This separate patent was also granted.

ON SEPTEMBER 30, 1948, Waksman received a royalty check for $44,474, bringing his total so far to $187,283. In October, he sent Schatz another $500 check, which he accepted "but only as a loan." Waksman was delighted to hear from Schatz that things were going well at the Hopkins Marine Station, and offered another check before the end of the year. The check arrived a month later—with an assignment form for the patent for South America. Waksman asked Schatz to treat the checks as "outright payments" to help him obtain advanced training in microbiology.

Schatz signed the patent form, but this time politely returned the check, saying that his financial situation was now "quite satisfactory" and he therefore could not bring himself "to impose upon you any further." He still had two hundred dollars from the first check, and his savings account was at eight hundred dollars. Vivian was still helping at the lab, earning about fifteen dollars a week. "We are living comfortably and are by no means stinting or denying ourselves."

Schatz added, "To be perfectly honest with you, I simply would not know what to do with the money if I had it, and I sincerely hope that I shall never have to spend any time attempting to solve such a problem myself." He and Vivian were neither smokers nor drinkers, and they did not go out much. They liked seeing their friends, music, reading, canoeing, and hiking. "We value most things that cannot be purchased with money." He thanked Waksman for his "friendship, confidence, and encouragement . . . things worth more than material aspects."

What happened next would trigger a sudden and explosive break in their relationship.

Waksman replied immediately. He "regretted" that Schatz had returned the check, explaining that the Rutgers Foundation had given him "a certain sum of money" and he felt that Schatz should be compensated for assigning the streptomycin patents to the foundation. There would probably be two or three more assignments. He asked Schatz to reconsider his initial rejection of the check, as otherwise, he wrote, he would have to pay his own income tax on the money.

Schatz was shocked. Waksman wanted him to pay income tax on funds that Waksman had been secretly receiving into his personal account from streptomycin royalties. But if the five hundred dollars was from strepto-mycin royalties, he was certainly entitled to it. So he decided to accept the check and argue about the tax matter later. "As to the check," he wrote, "I will now be glad to accept it if you will be good enough to send it to me again." He had not known the "source of the funds" and had not "wanted to be a ball and chain around your neck, nor to take funds that might be better used in your own Department at Rutgers."

Waksman rewrote the check and asked Schatz again to make sure to report all three checks on his income tax. He also asked for Schatz's signa-ture on two more patent assignments—for Canada and New Zealand. This time, he offered one hundred dollars for the New Zealand patent.

Now Schatz was totally confused. He had assumed the checks were a gift from Waksman, and the total sum, as a gift, was not taxable. Where the money was coming from was a mystery to him.

PART III · The Challenge

14 · The Letter

SCHATZ HAD ALWAYS BEEN EMBARRASSED THAT he could not explain the money side of his discovery to his colleagues, like Doris Jones, or his family, especially Uncle Joe. He had planned to ask Waksman about it before he left for California, but there had been no time for a meeting. After Waksman had sent him the second and third checks, he had written the Rutgers Foundation, but he had never received a reply.

It was time to demand answers. Schatz was now twenty-nine years old. Vivian was about to give birth to their first child. But what would be gained by pressing such a demand? More recognition for his discovery? A share of the spoils? He had told Waksman that these things didn't matter to him and that the life he had chosen to lead with Vivian was simple; he had no need for extra funds. More important than funds, however, was Waksman's help in passing on news about job openings and writing recommendations. Upsetting his professor by asking awkward questions could put this relationship at risk.

Much as he feared the possible consequences, though, one weekend at the end of January 1949, Schatz sat in his laboratory at the Hopkins Marine Station and wrote a seven-page letter to Waksman. In the past, he had always confided in Waksman about everything, in both his professional and his personal life; it had seemed a natural thing to do. He had sought Waksman's wisdom without question. But this time it was different. He was so nervous that the letter was repetitive, the words at times awkward.

He thanked Waksman for the notice of another job opportunity and reported on his latest research project. Then he switched abruptly to "several matters" that had been making him "more and more uneasy." He begged Waksman's "indulgence" for what was going to be "a rather lengthy letter and the imposition" that it would take on his time.

"I earnestly hope you will not feel . . . that I have stopped trusting in you . . . You once told me that in certain respects you have been as a father to me. Since that is true, it is in this sense, therefore, that I am writing to you as a son . . . I so very much want our friendship to continue that I am giving voice to what might otherwise become matters of an upsetting nature," Schatz wrote.

By signing so many papers, he said, he had "been slowly signing myself out of streptomycin completely." He had done so willingly because Waksman had told him often "how easily streptomycin might slip through our fingers and we then become only briefly-mentioned names in an ever-lengthening bibliography. To a large extent that is exactly what has happened to me."

He did not regret this "slow but progressive dissociation from the discovery of streptomycin." It was enough for him that it had "alleviated human suffering," he said.

Then Schatz brought Waksman back to the day at Rutgers in January 1945, when they had both sworn as "co-inventors" on the U.S. patent application. "You afterwards shook hands with me and stated that we were partners in streptomycin but that neither of us would profit financially from the discovery."

But his friends, including "laymen, attorneys, and professionals (microbiologists, biochemists, and the like)," had bombarded him with questions about the streptomycin patent. Who had received the royalties? What was the Rutgers Research and Endowment Foundation? Since he knew nothing about the foundation, he "felt like a damn fool, plainly and simply like a damn fool."

He didn't even know whether the streptomycin patent had been granted, or why Betty Bugie had been excluded from the patent, although her name was on the original paper announcing the discovery. "I believe you will understand how confused and perplexed I have become," he wrote.

As to the discovery itself, Schatz reminded Waksman, "You yourself know, better than anyone else, to what a considerable extent my own personal efforts are responsible for the discovery. You yourself told me six years ago that was why you felt it only fair and just that I be accorded senior authorship on the first paper announcing the discovery. You know the long hours and hard work I put in. You will recall on how many mornings you came to the laboratory at 8:30 or so and found the autoclave still very hot from my having worked through the night.

"All data of the original experiments of *A. griseus* and streptomycin are in my own handwriting in my laboratory notebook, which is now in your possession. I therefore feel, in all modesty, that streptomycin is, to a large and considerable extent, the fruit of my labors."

His anguish had become overwhelming, and he could not continue to sign over "unknown amounts of monies to which I am otherwise legally entitled to an organization about which I know practically nothing." And he would be reconsidering those documents he had already signed— unless he got satisfactory answers to a long list of thirteen questions, which he laid out one by one. They included questions about the foundation, the royalties, the foreign patents, and whether anyone had benefited personally. He felt he had a "lawful basis" for these requests because he was the senior author on the first streptomycin paper; such a request was "morally justified" in view of his work on streptomycin.

In case Waksman should read his letter as a threat, Schatz hastened to add, "I want you to know that I have, up to the present date of this letter, never approached anyone for advice on these matters. Until now, any suggestions and opinions which have been directed to me by others have not only been entirely unsolicited, but have even been discouraged by me." He was apparently referring to his family, and especially to Uncle Joe.

"I earnestly hope that you will be good enough to help settle these matters to our mutual satisfaction in the shortest possible time," he continued, "as unilateral action on my part would be not only difficult but embarrassing."

He had not yet decided what course of action to take. "To a large measure that will depend on your reply to this letter . . . But regardless of what will transpire, I can promise that I shall be true to my own convictions. Sincerely, Albert Schatz."

WAKSMAN'S HARD-EARNED, COMFORTABLE world, the personal fame and fortune he had amassed for the discovery of streptomycin, suddenly looked fragile. The question of who had discovered the drug, a matter he had thought successfully buried, had suddenly struggled back to life. His secret about the royalties might now be revealed, not merely to Schatz but to the world. How the revelation would be received at Rutgers, or at the National Academy of Sciences, or the other halls of high science in Europe, or the universities and colleges and hospitals that had rewarded and praised him for such a wonderful medical discovery was no doubt on his mind, as were his prospects for a Nobel prize.

Waksman's reply was short and angry. "To say that I was amazed to read [your letter] is to put it quite mildly," he began. "I can assure you that it caused me considerable pain. I thought that this whole matter had been settled and I was hoping that it would not come up again."

Waksman was apparently referring to the confrontation between the two men in the spring of 1946, when they had assigned the patent to the Rutgers Foundation. Drawing a distinction between the isolation of the microbe and turning it into a drug, Waksman wrote, "You know very well that you had nothing whatever to do with the practical development of streptomycin . . . You assured me of that yourself when you were about to leave New Brunswick in the middle of 1946. [He was referring to the letter dictated by Waksman to his secretary.]" He said Schatz had been given "all the credit that any student can ever hope to obtain." Schatz "knew quite well" that the methods for streptomycin's isolation had been worked out for streptothricin, and even the name *streptomycin* had existed before Schatz's return from the army.

Waksman insisted that Schatz's contribution to streptomycin consisted "only of the isolation of one of the two cultures" and in helping to isolate the crude drug. This was only a "very small part" of the development of streptomycin.

Waksman warned Schatz that this "whole matter" would now be referred to the attorney of the Rutgers Foundation. He was sending Schatz's letter to Russell Watson, attorney of the Rutgers Research and Endowment Foundation. "If I write to you again it will be in a far more formal manner than this letter."

And yet, angry as he was, Waksman appealed to Schatz, requesting that they settle their differences without the glare of publicity. He hoped Schatz would "reconsider this whole situation before it is too late. You have made a good beginning as a promising scientist, you have a great future before you, and you cannot afford to ruin it."

CLEARLY, WAKSMAN HAD never anticipated the extent of Schatz's rebellious mood. The other patents, for actinomycin and streptothricin, had been assigned by Boyd Woodruff without a murmur. But then, Waksman had arranged for a good job for Woodruff at Merck. And Woodruff had been a great success at the company. His temperament was different from Schatz's. He liked being part of a large organization. He was conformist. Schatz was the opposite. He liked to work on his own; he disliked "pressure from above," as he had once put it.

If Waksman had waited for a day or two, Russell Watson might have tempered his reply. Schatz's letter had presented Waksman with a stark choice: challenge Schatz or accommodate him. He could have given Schatz information on his deals with Rutgers and with Merck and acknowledged that Schatz was indeed the codiscoverer of streptomycin. And then, because he was basically in charge of the foundation's affairs—its staff never did anything without his approval—he could have arranged to give Schatz either a fellowship from the royalty fund or a share in the royalties. The sums required for such an accommodation would have been tiny in comparison with the amounts coming into the university coffers and his own pocket.

As for the five-hundred-dollar checks, Waksman could have said that the patent had been granted the previous September and further explained that the funds were only now able to be distributed. Under such a strategy, Schatz would have had no grounds for the legal action he was clearly contemplating.

Waksman was not prepared to take these steps. In his view, his personal deals with Merck and the Rutgers Foundation were none of Schatz's business. Selman Waksman was a professor of the old school. He was the master. His students were his apprentices. Instead of acknowledging Schatz's concerns, he scolded him.

But Schatz was no longer a callow researcher; he had broadened his scientific horizons as well as his understanding of human nature. In the three short years since he'd left Rutgers, he had dealt with life in other laboratories, under the direction of other bosses. As a civil servant in the Albany state laboratory, he had experienced the incompetence of local government bureaucracy. In Manhattan, he had been a researcher for a private nonprofit group and witnessed what he had seen as wasted effort and wasted funds. In California, he had encountered a brilliant academic community that thrived on the sparkling intellect and generous nature of its charismatic leader, C. B. van Niel.

Perhaps more important than the work experience, though, was the active support from his family, and from Vivian. Schatz had found a strong-willed woman in Vivian Rosenfeld. Vivian had an acute sense of right and wrong, of social justice and civil rights. She had witnessed how hard her husband had worked in his basement laboratory before successfully isolating streptomycin, and she was determined to see to it that he was treated properly and that his achievements were fully respected. In Albany, she had experienced the long arm of the FBI and watched as other academics had fallen victim to the growing power of Senator Joseph McCarthy and his Un-American Activities Committee. She understood how academic records could be twisted and used for unsavory purposes.

Money was no longer a great concern since Schatz's new job at Brooklyn College, which would begin in the fall, offered a good salary, forty-five hundred dollars a year. All Schatz now sought was due recognition as co-discoverer. He wanted to be properly informed about the royalties, instead of waiting for handouts from Selman Waksman.

For WAKSMAN, NOW aged sixty-one, streptomycin was the greatest achievement of his scientific career. The discovery had transformed him from a successful professor of a barely acknowledged scientific discipline at a modest academic institution into one of the most famous biologists of his era. He had turned down lucrative job offers in industry and posts at other universities. He had used his personal wealth to enhance his own reputation and that of the Department of Soil Microbiology with frequent visits to Europe, as well as trips to Russia and South America. In the process, he had picked up scores of honors and awards, some of them adding considerably to his personal fortune. As a microbiologist, he had been a special adviser to the U.S. government on important military matters. During the war, he had consulted on the role bacteria play in the fouling of ships' bottoms that reduces speed. He also looked into the bacterial action in steel corrosion. And he was involved in finding ways of reducing damage by molds and fungi to optical, electrical, and other military equipment. He was also involved in finding new sources of agar. The most plentiful source had been a red algae common in eastern Asia, and before the war most of it came from Japan. In these matters, he had found a place for himself at the Woods Hole Oceanographic Institution, where he spent his summers among colleagues from Harvard and MIT.

From his cramped office on the third floor of the Administration Building, Waksman ran his own antibiotic empire. He had personally arranged patents for streptomycin in several foreign countries. The Rutgers administration trusted him to make deals with some of the leading industrialists in the new age of antibiotics and with the increasingly powerful businesses producing pharmaceuticals.

Moreover, as streptomycin had gathered credibility as a cure for TB, Waksman had now been nominated for a Nobel prize four years in a row. He knew he was being considered; he had done his best on his European tours to make his presence felt at the Nobel "court" of the Royal Caroline Institute, in Stockholm. Having Albert Schatz loudly acclaimed as the co-discoverer of streptomycin could interfere with his chances.

If the Nobel Committee for Physiology or Medicine considered that the discovery of streptomycin was indeed worthy of the prize, then it would also have to consider the discovery as a whole, the isolation of the microbe and its development into a drug, as it had done with penicillin. Fleming had discovered penicillin, but Florey and Chain had turned Fleming's accidental observation into the world's first antibiotic. The committee had recognized all three scientists, the maximum number allowed for the prize, according to Alfred Nobel's eccentric will.

The streptomycin field was even more crowded. At the Mayo Clinic, Feldman and Hinshaw had shown the drug's effectiveness with their crucial guinea pig tests. Company chemists at Merck had worked on the extraction, purification, and chemical structure of streptomycin. And now, seven years later, Albert Schatz was pressing his claim to being the Fleming of streptomycin. Half a dozen people besides Waksman were possible prizewinners. The committee could decide that was too many and look elsewhere.

Waksman decided to fight and hope that the young Schatz would give in and settle the case out of court. Surely Schatz's claims would be crushed by the weight of Waksman's rank, his academic achievement, his honors and his awards, and by the superior firepower of his friends and colleagues, of Merck and its teams of lawyers, of the Rutgers PR Department. His former apprentice would be forced to retreat.

ELEVEN DAYS AFTER his first letter, Waksman wrote to Schatz again. Six pages long, this second letter laid out his counterattack. It was based on Waksman's parable of the chicken—now specifically labeling Schatz as a tool, a mere pair of hands. It was he, Waksman, who had identified the two streptomycin-producing cultures as *A. griseus*, a microbe that he had first isolated twenty-eight years earlier. One of these cultures had come from Doris Jones, the other had been found by Schatz. It was he, Waksman, who had ordered a detailed investigation of the two cultures, and his assistants Elizabeth Bugie and Christine Reilly had helped in those experiments. Thus, Schatz was "one of many cogs in a great wheel" and had played no part in later investigations.

"How dare you now present yourself as so innocent of what has transpired when you know full well that you had nothing to do with the practical development of streptomycin and were not entitled to any special consideration," he wrote. He "emphatically" denied that Schatz had any special rights to streptomycin, or that he had ever "suggested or believed" that he had any such rights, "or that you ever thought or mentioned to me that you had such rights."

Finally Waksman exploded, "What do you know of the headaches, of the sleepless nights, of the energy, spent to put the antibiotic across?"

If Schatz had attached any significance to the fact that Waksman had put Schatz's name first on the two key papers announcing streptomycin, then he was being naive, Waksman implied. In most other universities, he wrote, Schatz's name "would have probably been mentioned in a footnote, or at the end of the paper . . . I was proud of your abilities and attainments . . . In your case I was as generous as any professor could be expected to be."

As to Schatz's name on the patent, Waksman wrote that it was only there to show that Schatz had helped in the discovery. The Merck lawyer had "insisted" that Schatz's name be left out and that only his name be used. But he had "preferred" Schatz's name be included.

15 · Choose a Lawyer

EXPECTING A LAWSUIT, WAKSMAN LAUNCHED A furious and some-times wacky campaign to belittle Schatz's contribution to the discovery. He would seek evidence that Schatz was unstable and, worse, that as a laboratory researcher on streptomycin he might have doctored his notebooks to give himself greater credit than he was due. Waksman's preparation of his defense took on an air of desperation, hardly befitting a university professor and certainly not in the amiable, fatherly image that Waksman had created for his apprentices.

In the vicious confrontation that followed, rank clearly mattered, but so did class. Schatz and his family came from a different social stratum than Waksman, the kind of class distinction that was well drawn in the Russia they had left behind, and that still permeated their lives in the New World. Compared with Waksman, Schatz's father was not a success. He was an itinerant housepainter and a dirt farmer scraping a living in the depleted soils of Connecticut, and Schatz's mother had been a bakery shop assistant who had never been to high school. They were not like the educated and well-read family of Fradia London Waksman, the proud ma-triarch of Novaya Priluka, and her husband, Jacob Waksman, the owner of property in nearby Vinnitsa.

In preparing his counterattack in the spring of 1949, Waksman relied on the expert advice of Russell Watson, the sharp-witted lawyer for the Rutgers Foundation, and the media-savvy spin doctors of the Rutgers PR Department, always eager to defend their famous faculty member.

Watson's first concern was the question of the royalty payments. To lessen the impact, Waksman immediately volunteered to take a pay cut, slicing his royalty check in half, from 20 to 10 percent of whatever Rutgers received. In order to portray Waksman as a giver rather than a taker, the PR Department announced plans, which had been brewing for some months, for a new Institute of Microbiology to be built at Rutgers. According to the PR Department, it would be funded with a "gift" from Waksman of a million dollars. But this was a PR stunt. Waksman had not earned a million dollars in royalties from streptomycin to give away, and he had assigned the patent to the Rutgers Foundation; the rest of the royalty earnings belonged to Rutgers, not him.

At the same time, Watson and Waksman launched a series of bizarre attempts to discredit Schatz, and made a crude bid to buy his silence. In the years to come, Waksman would wonder how he could have avoided the unpleasantness that was about to be unleashed. For now, he was determined to protect his reputation, and his fortune, at all costs.

At Watson's request, Waksman drafted a memo outlining why he, and not "Mr. Schatz," should get the credit for the discovery of streptomycin. (He deliberately did not grant him the Ph.D. honorific of "Dr.") The memo repeated much of what he had laid out in his letter to Schatz. He had been the director of the research lab. He had always given his graduate students—of whom Schatz had been only one—directions as to how to proceed once an organism had been isolated. They were his tools, he stressed, not yet fellow scientists.

To reinforce his claim, Waksman now added a list of the occurrences when streptomycin was announced immediately after the discovery—"without the name of Schatz." These included a public announcement from his lecture in New York on November 16, 1944, when he discussed the great possibilities of streptomycin; his first radio address on streptomycin, given in 1944; the first broad summary of streptomycin, published in 1945; and the radio address telling the story of streptomycin, "where you find outlined the emphatic points." The name of Albert Schatz did not appear in connection with any of these special events, he triumphantly declared in the memo to Watson. Then he attached "Mr. Schatz's PhD thesis," which, he noted sarcastically, "must be read in the light of the general policy under which the graduate students submitting theses from our department are permitted to use data obtained by other students." In other words, Albert Schatz's thesis was not all his own work.

WAKSMAN THEN SOUGHT the support of former students who had worked for him during the discovery of streptomycin. On March 14, Waksman held a conference of four former graduates in his office. Among those present were Elizabeth Bugie and Christine Reilly, who had worked in the upstairs lab in 1943 when Schatz was in the basement. The meeting was written up by Sam Epstein, the author of the 1946 Rutgers-sponsored book on streptomycin, *Miracles from Microbes*, and who was now a paid consultant to Waksman's legal team. According to Epstein's account of the meeting, all four former graduates agreed that

> Schatz made no unique contribution to streptomycin because (1) he was, like other members of the staff, carrying out Dr. Waksman's directions; (2) doing no independent work; (3) the part done by Schatz could have been done by any of the other workers, had they been assigned to his task—as he could have done their work. In other words, the various assignments presented no special problems involving special skill or knowledge. Reilly and Bugie also agreed that Schatz had had closer contact with Waksman than other members of the staff, Bugie claiming Schatz's (four times a day) trips up to Dr. Waksman's office from the basement laboratory becoming something of a joke among members of the department.

Next, Waksman asked Schatz's former employers about his "personality" and work habits. Waksman already knew that Schatz had had difficulty in those first years trying to find the right work for himself, because Schatz had written him several times. Those early misfits—the way Schatz and his job had not matched and his tendency to be a loner—could be very useful to Waksman now in portraying Schatz as somewhat unbalanced and even unreliable.

He asked Dr. Chester Stock, head of the chemotherapy research division at Sloan-Kettering, for "a confidential opinion of Dr. Schatz's personality, especially his relations with the other workers both in your group and the other divisions of the Institute during his stay at Sloan-Kettering."

Without mentioning his confrontation with Schatz, he wrote, "Certain matters have come up recently which have made me quite suspicious of some of his activities, both here and in the other laboratories where he has been since he left us in 1946."

Waksman said he was "trying to collect, therefore, information concerning his associations with workers in other Institutions." He asked Dr. Stock for "as frank an expression as you can concerning him, his work and other matters that may have a bearing upon his personality and his ability to get along with other people. All this information will be kept highly confidential."

But Waksman had no intention of keeping the information he gleaned confidential. He was preparing to use it in court.

Dr. Stock was ready, even eager, to oblige. By return mail, he wrote that if Schatz had not left to go to C. B. van Niel's, "we would have found it necessary to request him to leave," but he did not explain why.

Stock admitted that he had allowed Schatz to spend from "one fourth to one third of his time on research of his own choosing." But that meant Schatz "worked upon everything but the problem he came here to do . . . He processed [sic] with the requested studies in an indifferent fashion and allowed himself to be easily stopped by the problems encountered. Some of the problems are not easily handled, if at all, but Dr. Schatz failed to show initiative and enthusiasm in trying to overcome the difficulties. I believe it was due to his great interest in his other studies."

Yet Waksman already knew that Schatz had felt underemployed at Sloan-Kettering, mostly because the labs had not been ready. Out of boredom, he had taken Russian lessons at Columbia. They had discussed this in letters, and indeed, Waksman had advised him to leave and go to California and van Niel.

As to Schatz's personality, Dr. Stock said that Schatz totally lacked a "team spirit." In seminars and other discussions his questions and comments were seldom "made in a friendly spirit." His attitude at times might well be summarized as "anti-social," and he had "rapid changes of interest." This was "possibly" merely an indication of a keen, active mind, "but I have wondered if it might not be indicative of a certain instability which is also reflected in his somewhat anti-social attitude and actions."

Waksman also contacted Gilbert Dalldorf, the director of the Department of Health in Albany and Schatz's first employer after Rutgers. Dalldorf wrote, "I personally was very fond of Dr. Schatz and was disappointed that he left. He was interested in opportunities for advanced study." However, he added that "it was only after he left that I learned that he had been malicious toward one of the members of staff in a rather irresponsible manner. Perhaps this was an aberration that will never again repeat itself." There was no further explanation.

FOR HIS AGE and relative inexperience, Schatz, with the encouragement of his parents and the help of Uncle Joe, proved to be a surprisingly formidable adversary. His strength came from one main source: He knew that truth was on his side. He was not as concerned as Waksman by the prospect of a lawsuit; he had no money to lose. He took on his new mission with the same determination he had displayed when he faced the daunting task of finding the microbe that could kill TB.

His first move was to inform his former colleague Doris Jones. She was doing her Ph.D. at Berkeley, about a hundred miles up the coast from Albert and Vivian in Pacific Grove. Schatz sent her a note about his exchange of letters with Waksman, and she immediately reassured him of her total support. There were "no skeletons hidden in any of her closets concerning his part in the discovery of streptomycin," she said. It was the work of Schatz and Schatz alone, toiling in the basement of the Soil Microbiology Department.

"At no time," she wrote, "have I ever regarded myself as being in any way instrumental to your isolation of the two organisms . . . If Dr. Waksman now claims I had as much to do with it as you or even Chris [Christine Reilly] or Betty [Bugie], he is entirely mistaken . . . I don't know why all this to do over who did what at this late date, for you know as well as I the story behind the story.

"I felt as strongly as any that your credit—at least in so far as the isolation and initial studies goes—was entirely discouraged and for reasons Dr. Waksman and only Dr. Waksman can answer."

There had been many times, she said, when she had wanted to "burst out to people and tell them what I knew had happened—from my viewpoint—but I have for the most part held my tongue because I knew that it would serve no purpose." And she knew how bad Schatz had felt about the lack of credit from Waksman. "I can appreciate and sympathize with your disappointment in seeing so much of your work incorporated into the glory of Dr. Waksman and your chances of gaining recognition falling by the wayside." It must have been especially hard because Schatz had been, as she put it, "more or less of an idol-worshipper" when it came to Selman Waksman. "I can see how that faith you had in him must have been terrifically shattered . . . [You] expected that right would be done because it was promised."

She had found it difficult at first to see how Waksman could take the credit without a guilty conscience, but "as the years rolled by," she had accepted his contention that he had been pushing and directing and consulting in many ways with many people to make the development possible.

Oddly, Waksman was courting her now, she revealed. After having had no help from him in the way of grants for three years, she had received a letter asking if she would be interested in another position at Rutgers. She saw the offer as an attempt to attract her support in the coming legal battle with Schatz, and she had rejected it.

A second ally turned out to be Seymour Hutner of Haskins Laboratories in New York. Schatz had spent time with him while he was at Sloan-Kettering. When Schatz told him about never having been paid for signing over the patents, Hutner replied, "There's nothing very surprising in your imbroglio with Waksman, except that I thought he was cleverer than to get himself on record with such an arrant swindle." Hutner warned, "When writing to the Rutgers boys, a good attitude might be a frigid detachment—you play into the hands of people like that if you lose your temper." Hutner advised Schatz to be his "old self . . . contumacious, rambunctious, irreverent, sacrilegious, heretical. Show me."

IN THE LATE spring of 1949, the Rutgers PR Department launched its own offensive. In a gesture of astonishing generosity, Professor Waksman, it announced, had "turned over the patent for streptomycin for the establishment of a new Institute of Microbiology at Rutgers." The headlines blared: "Waksman Rejects Chance for Wealth" and "Streptomycin Income Goes to Rutgers for New Institute." The stories came from the Rutgers press release, and they were wrong. Certainly, the new institute was Waksman's idea, but the money was to come principally from the streptomycin royalties to the Rutgers Foundation, not out of the percentage of the royalties in Waksman's pocket. Such details didn't bother the alumni association, however. The class of 1948 received a solicitation for a "fiver or better." "Remember, Waksman gave a million!" it read. "What are you going to give?"

In its press release, Rutgers did not mention Schatz, but the *Passaic Herald-News*, Schatz's hometown paper, never missed an opportunity to promote its local hero. "A former Passaic man participated in the discovery that is making the new institute possible," it noted. He was "Dr. Albert Schatz, who worked at Rutgers under Dr. Waksman and who is recognized as the co-discoverer of streptomycin." In the end, the Rutgers promotion backfired. When Schatz read about the "one million dollar gift," not knowing any better, he took it seriously. For him this was the first indication of the size of the streptomycin royalties gleaned by Waksman behind his back. Schatz's father, Julius, cabled his son in Pacific Grove: "Choose a lawyer."

Uncle Joe agreed, but he also had another idea. He enlisted the help of a friend who ran a Manhattan public relations company, M. D. Bromberg and Associates. The owner, Max Bromberg, "represented international companies," and he was more than a match for the Rutgers PR team.

Bromberg sent identical letters to several current and former members of Waksman's staff at the Department of Soil Microbiology. The letter began, "I have recently been in correspondence with a Dr. Albert Schatz at the Hopkins Marine Station, Pacific Grove, California, concerning his writing for publication in a popular periodical an article on 'The Discovery of Streptomycin.'"

Dr. Schatz "has assured us that he is one of the discoverers of streptomycin." He had sent Bromberg a number of publications to "substantiate his claim." These included the original 1944 scientific papers and a copy of U.S. Patent No. 2,449,866, awarded in September 1948 and naming Schatz and Waksman as codiscoverers.

"Being laymen with respect to the field of science," Bromberg continued, "we are unable to evaluate Dr. Schatz's claim that he is one of the discoverers of streptomycin in view of the confusing situation that our radio and press mention only Dr. Waksman as the discoverer.

"In order that we may arrive at a fair decision regarding this matter, we should appreciate it if you would be kind enough to give us your opinion

as to: (1) exactly what role did Dr. Schatz play in the discovery of streptomycin and also (2) whether you consider him to be one of the discoverers
of this drug."

Those who received the Bromberg letter took it seriously. One of the
first to reply was a former associate professor of plant pathology at Rutgers from 1938 to 1947, Dr. P. P. Pirone. He was now at the New York Bo-
tanical Garden, in the Bronx. Pirone wrote, "I suppose (as is the case with
most research institutions), Dr. Waksman put him to work on the particular organism which later was found to produce streptomycin. The fact re-
mains, however, that Dr. Schatz should be credited, at least, as co-discoverer
of this chemical." Pirone cited as evidence the streptomycin patent. Additional proof was Schatz's doctoral thesis, for which Pirone had been one
of the final examiners. Pirone recalled, correctly, that some earlier, popular
stories had given Schatz credit but that "as time rolled on and as the discovery assumed greater and greater importance, his name was dropped
from all radio and news releases."

"I have always felt that Dr. Schatz was treated very unfairly in the whole
situation," he concluded, and he hoped that "some day proper credit" would
be given to him. "He expressed his disappointment to me personally several
times and I always told him 'the truth would out' some day."

Doris Jones also received the Bromberg letter. It was "an absolute fact,"
she said in her reply, that Schatz had been "solely responsible for the isolation of the first two streptomycin-producing strains . . . One of these he
obtained from the soil; the other from a plate prepared by myself from a
swab of a chicken's throat."

She said that Schatz had "tested countless cultures before by a lucky
stroke of chance he found one which looked promising. I say 'chance' for
the soil and other materials contain great numbers of microbes, any of
which may or may not possess the properties essential for the production
of an effective antibiotic agent." As far as Jones was concerned, "there would
have been no streptomycin without a Dr. Schatz—at least at that time.
The fact that relatively few other active strains have been found since
then give active support to the element of chance in the isolation of the
culture."

The techniques used by Dr. Schatz were "not original," she said. "They
had been known for years." They were "but a tool to the search, just as a

Geiger counter might be to the discovery of uranium." It was due to Schatz's "zealous persistence," his attempts "to test as many cultures as humanly possible . . . that he finally pulled *S. griseus* out of the microbial grab-bag." She added, "Had it not been for him, my plate containing one of the first two strains might have been tossed away." Schatz was so good at the lab-bench research that he "carried on independently" of the teachers. His contribution was "definitely fundamental" to the discovery, and she personally felt that "the radio and press—and even scientific publications—had neglected to give him due credit."

Others who received the Bromberg letter strove for balance, like Kent Wight, a former Waksman graduate student who had been in the Department of Soil Microbiology at the time of the discovery. He said that Schatz "is one of the discoverers of streptomycin," but he did not wish to minimize the part Dr. Waksman had played. "Each person who had worked with Waksman over the years had made some contribution." In another letter, Boyd Woodruff suggested that Bromberg should contact Waksman directly. "He is always happy to give credit to his associates on his research projects." The dean of the Rutgers College of Agriculture, Dr. William Martin, said that Schatz was indeed one of the discoverers, but the "original thinking" behind the discovery had been contributed by Dr. Waksman, not by Dr. Schatz. He suggested that any article by Schatz should be peer reviewed before publication.

WITHIN A FEW days, Waksman began receiving his own copies of the Bromberg letter, forwarded by loyal staff members. He was furious. He wrote on top of one, "Malicious, I might say, if not sinister." But Waksman then took the dirty game to a new level. From seeing previous letters signed by Schatz as Dr. J. J. Martin, and assuming that "Dr. Martin" was a Schatz family member, Waksman was sure the Bromberg letter was a trick, but who were Bromberg and Associates?

In a bizarre move, half in jest apparently, Waksman and Russell Watson formed the "W and W Sleuthing Agency." Its goal was to "discover and prosecute all those malevolent persons who misuse their presence at the Institution for private and undesirable purposes." The "Institution" the Rutgers college farm. The two officials of the agency were named as Russell

E. Watson, attorney and chief prosecutor, and Selman A. Waksman, scientific sleuth. Its first report was titled "Information Gathered Concerning Dr. J.J. Martin."

"His name was J.J. March, but he had it legally changed to Martin," the report stated.

> He claims to be a cousin of A. Schatz, a co-discoverer of streptomycin, and has been known to express that opinion to several people. He has worked for an Advertising Dentist Concern . . . This is believed to be the poorest type of dentistry. Someone vaguely remembers that there was a dental technician that solicited patients for Dr. Martin saying that Martin did the dental work and he the technical work. It was also known that this technician has handed out his cards in such places as barrooms.
>
> A number of dentists in Passaic County were asked for an opinion of Dr. Martin. They were rather cagey in expressing such an opinion. They said they knew him as a fellow practitioner, but none had anything complimentary to say about him. They have no regard for him at all in the dental profession.

In Report No. 2, a "search has been made" of the premises (of Bromberg and Associates) and the investigator was told that they were engaged in "some sort of advertising, mostly in foreign publications." One office was a credit equipment corporation, the second a realtor. "No further information could be obtained concerning the above group."

Report No. 3 could "not find the company in any Directory or Publishers Weekly or Dunn and Bradstreet. Various banks have also been checked. They have no credit reference of any kind."

IN SEPTEMBER, SCHATZ came to New York to take up his new appointment as an assistant professor at Brooklyn College, and Bromberg had found him a Manhattan lawyer named Louis Libert. Schatz had no need to worry about legal fees, Libert reassured him. The case would be conducted on a contingency: Schatz would pay him only if he won.

Libert quickly discovered, from Waksman's federal income tax returns, that he had a personal income of $124,000 in 1948. The only conceivable source for the majority of it was a payment from streptomycin royalties, Libert concluded.

At the end of November, Russell Watson agreed to an extraordinary meeting at the Union League Club, on East Thirty-seventh Street and Park Avenue in Manhattan. Waksman and Watson were on one side of the table, Libert and Schatz on the other. Schatz felt that they might want to settle their differences, as he was keen to do, but Watson quickly killed any chance of compromise. He offered Schatz one thousand dollars—a "nuisance value" payoff for his trouble in putting his signature on patents for foreign countries. The sum was not what Schatz and Libert had in mind. For all they knew, the foreign patents were earning somebody, or some corporation, millions of dollars. "The party adjourned to a luncheonette, and broke up," Schatz's lawyers recorded.

The battle was on, and Libert felt he needed help. He contacted his friend Jerome Eisenberg, a quick-witted New Jersey trial lawyer. Eisenberg had made a name for himself mostly in tough civil cases—defamation, contested wills, tax appeals, and torts. During the depression, he had listed as one of his achievements that in New Jersey he had foreclosed more homes than any other lawyer. He was a legal street fighter and just the man Schatz needed. He agreed to take the case on contingency.

Schatz told Seymour Hutner in New York, "The die is cast."

BUT ON THIS phony battleground, Waksman, with his superior arsenal, was to have the last word before he and Schatz were technically at war. At the end of October, the Royal Caroline Institute announced the winners of the 1949 Nobel Prize in Physiology or Medicine. The prize was given for advances in brain research, including the prefrontal lobotomy, then a last resort of schizophrenics and manic-depressives. The recipients were the Spanish neurosurgeon Dr. Antonio Moniz and the Swiss physiologist Dr. Walter Hess, a specialist in circulatory and nervous systems. But the scientist on the cover of the November 7 issue of *Time* magazine was Dr. Selman Waksman. And the headline on the article, inside, which ran over six pages in the Medicine section, was "The Healing Soil." It was a story about antibiotics, and it mentioned Alexander Fleming's penicillin and two new antibiotics by American researchers, but it concentrated on Waksman's "discovery" of streptomycin. Albert Schatz was not mentioned. The intent of the story seemed to be that Waksman should be next for the Nobel Prize.

The story began with a public relations lie, put out by Rutgers's PR Department. According to the story, Waksman had almost been fired at the start of World War Two when Rutgers was looking to make staff cuts. Why should it continue to pay a soil scientist who was "playing around with microbes?"—and being paid $4,620 a year to do it. "Fortunately for mankind," Dean William Martin had "saved Dr. Waksman from the ax," and it had paid off handsomely—streptomycin's "harvest of pennies" from the royalties had already brought Rutgers more than two million dollars, and more was on the way. Waksman's salary was ten thousand dollars, and he assured *Time*'s correspondent, "Rutgers won't let me starve." A footnote said that he might even "get a percentage of the gross take."

Waksman knew that the story of his almost being fired had been made up by an overeager Rutgers PR man, but he did nothing to correct it. He did write a letter to the editor of *Time*, though. The original story as published had not mentioned Schatz. Now, Waksman asked that "the names of the students most closely associated with the isolation of streptomycin in 1943" be recognized." He named them, in the following order, as Miss Doris Jones, Dr. Albert Schatz, Miss Elizabeth Bugie, and Dr. H. Christine Reilly. And *Time* published his letter in the November 28 issue. There was also a reader's letter that began, "I was delighted to see the face of Dr. Waksman peering from the cover of *Time*. The pictures of politicians, prizefighters, musicians, models etc are all right their small sphere; but the work of men like Dr. Waksman, which results in good for all mankind, regardless of race, creed or color, is of much greater importance."

16 · The Road to Court

JEROME EISENBERG LIKED THE SCHATZ CASE. As a trial lawyer, he was drawn to the "David and Goliath" aspect, as he crudely put it. Libert warned him that Schatz had no money, but Eisenberg agreed to take the case on contingency—if Schatz passed his legal litmus test. He wanted to evaluate Schatz's credibility, appearance, personal history, good faith, social life—no "extra-marital affairs"? It was the kind of thing he usually asked his clients. Most important, he wanted to make sure Schatz would not be "taken for a communist." Vivian had worked for civil rights and left-wing causes; although Schatz's sympathies lay in that direction, he had never joined any groups. In the face of demands that professors take loyalty oaths, Rutgers "blew with the wind. They were conservative when it was fashionable and liberal when the pendulum swung the other way," Hubert Lechevalier, one of Dr. Waksman's graduate students, later recalled. It paid to be prudent. If the Rutgers authorities had the slightest suspicion that Schatz was a "fellow traveler, red, pink or otherwise," Eisenberg said, the case might be lost on that count alone.

On February 13, 1950, Schatz spent twelve hours with Eisenberg in his Newark office. He gave the lawyer a quick course in how to find an antibiotic in a thimbleful of farmyard soil, and Eisenberg outlined the case he thought he could bring. He would charge Waksman with fraud and deceit for hiding the money he had received from the patent. And he would include the Rutgers Research and Endowment Foundation in the suit. At the end of the session, Eisenberg was satisfied that Schatz would make a "most credible and intelligent" witness. "His honesty was evident in his appearance and candor. He would answer questions truthfully and re-sponsively; he had no skeletons in his closet."

However, there were tricky legal issues that made this more compli-cated than Eisenberg's previous civil cases. For example, who had originally held the patent rights? Did Schatz have rights as a student in a college labo-ratory whose research was part of his course for a doctorate? Did Waks-man have rights as a member of the college faculty? If Schatz was paid a tiny wage, in addition to his stipend, by the college for looking after the plants in the greenhouse, not for his lab work, did the "shop rights" doctrine apply, giving rights to the employer? Was Schatz obligated to assign patent rights to the college that employed him?

Like a good trial lawyer, Eisenberg examined all angles, even the im-probable ones, such as the question of fraud perpetrated on the U.S. Pat-ent Office. If it could be proved that Schatz was the lone discoverer in patent law, and that Waksman had no part in it, then the patent itself might be void, a matter that would have to be decided in federal court. If a fed-eral court held the patent void, who would then be entitled to claim the royalties already paid to the Rutgers Research and Endowment Founda-tion? These royalties could not be returned to the purchasers of the drug because many of the consumers could not be found. Would the royalties remain with Rutgers? Would Merck, a licensee of the patent from Rutgers, be entitled to the return of any of the royalties?

Schatz was troubled by the discussion. Eisenberg judged him to be in a state of "moral torment" over whether he should be receiving royalties at all. And he could not, of course, even assure Schatz that, if the patent were to be found void, he would be entitled to any of the royalties. Despite as-surances of the contingency basis of the arrangement with Eisenberg, Schatz was still worried that he might be incurring expenses from his own pocket, which was nearly empty.

The threatened departure of Waksman for a European tour, which would put him out of reach for a few months, persuaded Eisenberg to draft a com-plaint without delay. It alleged that Schatz and Waksman were "co-discoverers," as the patent application had shown; that Schatz had conducted experiments that "led to the isolation and discovery" of streptomycin; that these experiments were "checked and confirmed" by Waksman, who "col-laborated in the further development of the drug"; that for some time there existed a "close personal relationship" between the student and his professor; and that on February 9, 1945, they applied together for a patent that Waksman said was to prevent others from gaining a monopoly, controlling the price of the new drug, and profiting from it. Then, "on or about May 3, 1946," Waksman asked Schatz to assign the patent to the Rutgers Foundation, and "by virtue of [Waksman's] power, position, and influence in the field of microbiology and the world of science, as contrasted with Albert's youth and inexperience," Waksman threatened that if Schatz did not sign over the rights, Waksman would "see to it" that Schatz would fail to gain employment in the scientific and professional field for which he had been trained and that no reputable institution would employ him.

As "co-discoverer," a fact admitted by William Martin, dean of the College of Agriculture, in his reply to the Bromberg letter, Schatz "was and is the owner of a half interest in the royalties from the patent," Eisenberg asserted. Waksman had represented himself "to be the sole discoverer" of the drug and "had reaped rewards" not shared by Schatz. Finally, and brashly, the complaint demanded for Schatz half of the payments made so far to his former mentor.

REPORTS OF THE suit appeared on March 10 in the *New York Daily News* and the *Newark Evening News* under the headline "Suit Accuses Waksman: Former Student Says He Discovered Streptomycin, Demands Accounting from Foundation." The next day, the story appeared in the *New York Times*, the *New York Herald Tribune*, and halfway across the nation in the *St. Louis Post-Dispatch*. In Schatz's hometown paper, the *Passaic Herald-News*, the story ran under the headline "Ex-Passaic Man Sues for Profit of Wonder Drug." Initially, Russell Watson said the claim was without merit and would be "vigorously contested." Contacted at his home in Brooklyn, Schatz declined to discuss the case.

In the *New York Times*, Watson pumped up his response, calling the suit "baseless and preposterous." Streptomycin was the result of nearly thirty years of "continuous, systematized" study by Dr. Waksman, he added. "Dr. Schatz was one of about twenty graduate students who aided Dr. Waksman from time to time in the study of this problem . . . He assisted in the prosecution of a definite part of a comprehensive plan utilizing technical procedures devised and directed by Dr. Waksman, based upon his many years of scientific research and experience in microbiology . . . The trial of

Dr. Schatz's unfounded action will conclusively demonstrate that his charge is false."

DORIS JONES WROTE Schatz as soon as she read the news. Again she pledged support, but she also warned of the risks.

"Dear Alberto," she wrote, "I can understand now why you haven't let the Western outposts know of your doings for the past few months. I certainly hope the whole thing turns out a great success—moral and otherwise. You know exactly how my sentiments, opinions etc., etc., go, and if I can be any help, Al, don't fail to call on me. I hope, if you still need them, the records are still on file where they ought to be."

She was willing to be a witness if needed, she said, adding, "I suppose you must consider you have an airtight case to attempt anything like a suit against the hierarchy of Rutgers." But she worried that somehow he might, as a result of the lawsuit, "get involved in commitments to others"—i.e., lawyers. "But then there's no need to worry about you because you are possessed of a great deal of common sense."

A few days later she wrote again. Schatz's suit had "certainly stirred up a tremendous amount of talk in the circles of science." She had spent several long hours explaining to one colleague how Schatz had "come to blows (legal) with Waksman." But she warned Schatz that the reception to the lawsuit was not good. She was beginning to hear Schatz's reputation being put on trial. "You are being judged quite arbitrarily by men who consider your actions to be one of a money-grabber. Of course, everyone has a right to do what he wants with his own life—and his own money, and you are the last person I can imagine who would be out to claim money for money's sake." Another of her colleagues had observed, "Doesn't he [Schatz] realize that whether he is right or wrong he has ruined his reputation as a scientist!" Jones added, "I'm sorry to hear that, too, Al."

Jones pressed the point, as a friend. "Whether you are right or wrong in the eyes of the law, people out here do not interpret your action as anything but an attempt to grab onto money," she wrote. "That, Al, is very unfortunate insofar as your professional career goes." As far as she knew from the newspapers, Waksman had given all his money to research.

"Despite what we know of Waksman's character," she continued, "of what happened along the pathways of streptomycin development, it does remain a fact that he assigned all the money to a foundation for the furtherance of science. That fact puts him in a good light with all the men in science—and you, on the contrary, are painted as the cheap charleton [sic] who is jumping on the bandwagon for $$."

Perhaps, she suggested, Schatz should make some statement to the effect that any money he received would be turned over to some scientific foundation. "You wouldn't look so bad."

AS EXPECTED, THE scientific establishment rallied behind the master, with several of Dr. Waksman's colleagues sending him effusive letters of support. Of course, they did not know that he had been secretly enriching himself.

One of the first to write was Dr. A. J. Goldforb, Waksman's good friend and the editor of the *Proceedings of the Society for Experimental Biology and Medicine*, which had published the first paper announcing the discovery of streptomycin in January 1944. He called the lawsuit "scandalous" and offered any help needed. Dr. Arthur Wright of Albany Medical College was "very sorry that Dr. Schatz . . . has debased himself by taking the action he has." And Dr. Stanley Thomas of Lehigh University, in Bethlehem, Pennsylvania, was "shocked" to read of the suit. As an admirer of Waksman's, he was "convinced, without knowing any of the particulars, that it was this generosity on your part that made the present situation possible." In other words, he believed the publicity about the million-dollar gift.

William Steenken, head of The Trudeau Laboratory, in Saranac, New York, agreed with Waksman that Schatz "had no right to claim any part of the proceeds of the drug. *Since this money is not being used by you personally* [emphasis added], but is going to Rutgers where Doctor Schatz was only a graduate student working on a specific part of your overall program, I cannot understand how he could accept payment under the circumstances. To take such a stand is certainly unethical." In his summary of Steenken's letter that he sent to Russell Watson, Waksman omitted the key phrase "Since this money is not being used by you personally."

Watson was less impressed by these letters than his client. Sooner or later, Watson knew, Waksman would be forced to reveal the royalties he had been receiving. He wanted Waksman to draft a statement in preparation for that moment.

Waksman still hoped it would not be necessary to issue such a statement, but just in case, and as he was about to depart for his European tour, he sent Watson some "suggestions." First, he said, the statement should point out that since the beginning of the antibiotic hunt, in 1940, he had received "several highly lucrative offers" from American drug companies, but he had decided to stay at Rutgers "at the modest salary" the university paid. Second, he had "never expected nor did I want any additional compensation for my scientific work."

In "private conversations" with the Rutgers Comptroller, Waksman continued, "it was agreed that while most of the funds accruing from . . . patents should go to Rutgers University, a certain percentage" of such funds would be turned over to him for distribution as "I see fit." When it later became apparent that the sums accruing from the streptomycin patents might "become quite excessive . . . I took a voluntary, and quite appreciable, reduction in my compensation." In reality, Waksman had volunteered to cut his share only after he had received Schatz's letter of complaint in January 1949.

17 · Under Oath

WHEN JEROME EISENBERG ASKED TO DEPOSE Waksman in a pretrial hearing before he left for Europe, the judge agreed, but Waksman said he didn't have time. He was due at an important meeting of the World Health Organization in Geneva, to be followed by several other conferences, and it was not clear when he would be back. The judge was not impressed, and the court issued Waksman a subpoena.

Thus the stage was set for a pretrial confrontation between Eisenberg and Waksman, a rare opportunity for a lawyer to question a scientist about his work. Eisenberg was a seasoned trial lawyer, a fast learner of unfamiliar concepts and obscure, technical terms. Waksman was supremely confident that his superior knowledge of his science would see him through the interrogation. However, he would look back on the day as one of "seemingly endless questioning," while Eisenberg would see it as one of the most interesting depositions of his half century at the bar. The four-hour deposition began at 10 A.M. on March 25 at Russell Watson's office in New Brunswick. Present were Eisenberg, Schatz, Waksman, Watson, and their assistant, Samuel Epstein. It was a very tense session, the first of several that would involve various witnesses over the next five months.

Eisenberg began by quickly going over Waksman's background and then plunging into the main issue of who, exactly, had discovered streptomycin. Waksman's reply to this question set the tone of the meeting.

"The major part of the work on streptomycin was done with my own fingers," Waksman said. "No! No!" Schatz would later write in the margin of a copy of the deposition.

At the time of the discovery, Waksman maintained, he was spending 80 percent of his time in the laboratory. "I had several assistants," he said. "One of whom was Albert Schatz, and another was Elizabeth Bugie, and a third was H. Christine Reilly."

So, when it came to writing up the experiments and preparing the papers, Eisenberg asked, how much was Schatz involved?

"Very little," Waksman replied. Schatz's job was to keep records of his experiments, he said, and he produced a handful of tan cloth-covered notebooks, two of Schatz's and three of his own.

He would ask Schatz to prepare tables of the data from the notebooks, he said. "And he would prepare them exactly as I asked. He wouldn't do anything more than my secretary would do, or than any assistant to whom I would give specific instructions. He didn't do it of his own free will; never took initiative."

Eisenberg asked about Schatz's thesis, the one that showed how he had isolated the strains of *A. griseus* that produced streptomycin. Waksman said he couldn't remember its title—even though he was one of the examiners.

> E: You don't recall the subject of the thesis?
>
> W: No, it dealt largely with streptomycin, of course.
>
> E: Did it present original and creative work on his part?
>
> W: In a sense, yes, and in another sense, no. You see, the graduate students in my department—Mr. Watson, will you please let me make a simile?
>
> WATSON: Go ahead.
>
> W: Have you watched an orchestra perform and the conductor giving a cue to each musician? The musician himself plays only a very minor role, but together they produce a beautiful symphony ... In a sense that was original work, of course, but that work was as much a part of me and of my work as anything can be.
>
> E: Did you read his thesis?
>
> W: Yes, sir.
>
> E: Did you approve of it?

w: I did.

e: Did you make any changes in it?

w: Probably did, plenty.

Waksman then recounted how his lab had produced the first antibiotics—actinomycin and streptothricin—and then streptomycin.

w: We had all the methods worked out. We knew exactly what the culture was [*A. griseus*] because I named that culture in 1915. It produced streptomycin in 1915, but at that time we were of course not interested in antibiotics.

Schatz wrote in the margin, "It did not produce streptomycin."

Eisenberg returned to the actual discovery.

Only seven or eight weeks after Schatz had come back from the army in the summer of 1943, Waksman said, "we observed certain cultures. They looked as if they were active against Gram-negative bacteria . . . Since we had all the matters worked out for streptothricin, we at once continued further . . . All I needed was hands. The wash girl could have done exactly the same thing. She could have done exactly the same thing, if I put her on that problem. Schatz did nothing original, absolutely nothing original. He followed exact instructions that I gave. He would come up to see me three or four times a day. The girls in the laboratory would make fun about his running. To every little test he would come up to check with me that he is doing the right thing, and thus streptomycin came into being."

Schatz wrote in the margin, "My Ph.D. required original and creative work."

Eisenberg pressed the question.

e: You said, in talking about the experiments that led to the discovery of streptomycin, that "we observed cultures that looked good against the Gram-negative bacteria." Whom did you mean by "we."

w: I have always bent over backwards in giving as much credit to my assistants and graduate students as I possibly could. I even got so accustomed to it that I would use the word "we" in a general sense. That means if it happened to be Schatz, it was Schatz and I. If it happened to be Miss Bugie, it was she and I. If it happened to be Miss Reilly, it was Miss Reilly and I.

E: I meant what did you mean by "we" when you used it in the sense in which you testified earlier when you said, "We observed cultures that looked good against Gram-negative bacteria"?

W: In this particular case it was Schatz and I.

E: Now, you said that during 1943 there were other assistants of yours who were doing the same thing?

W: That's right ... There were my technical assistants who sometimes rendered higher service to some of the scientific assistants, but whose names never appear on scientific papers ... First: Miss Clara Wark [a lab assistant]. There was Dorothy Randolph [a lab assistant]; there was Mr. Cooper [Aldrage Cooper, a lab helper]. I don't remember his initial. There was Mr. Adams [who washed glassware and prepared media].

In his first public answer to Schatz's lawsuit, Russell Watson had said that Schatz was "one of about 20 graduate students" and assistants who had helped Dr. Waksman in the laboratory. Eisenberg now asked how many of the nineteen others had worked on streptomycin.

"Well, it depends entirely what you call worked," Waksman replied. "Is the one, Dr. Schatz, who was instructed to isolate the cultures and test them, is his work more important than the man who washes the dishes the cultures were made in, to the girl who prepares the media which is used in the preparation of those cultures, to the one who tests the material in animals, to the girl who grows the culture in the flask?"

Eisenberg pressed the question: What was the contribution of each of those nineteen to the isolation and discovery of streptomycin? Waksman replied that "probably the contribution of Boyd Woodruff, who assisted with the isolation of streptothricin ... was, oh, as much as twenty times that of Schatz," because he had helped in the development of procedures and ideas.

And what about the lab assistants and the glassware washers? Eisenberg asked. What was their contribution?

w: Oh, I suppose [on] streptomycin itself, there again, I suppose, Schatz was an important contributor. I don't mean to
say he has not contributed anything. Don't misunderstand
me. He was an important contributor, otherwise I would
not have put his name on these [17] papers. I would like
you to go to the universities of the country and show me
any university where a graduate student in the period of
three years has his name on 17 papers. The others, their
names may not have appeared on as many papers, but to
evaluate the contribution is rather difficult. Each one fits
into a certain, as I said, mosaic. Do you see what I mean? It
would be very difficult to say I have made 75 percent contribution, Schatz made 10 percent, Miss Bugie 5 percent,
Miss Reilly 5 percent; it would be very difficult.

Still, Eisenberg pressed the question: What contribution had the other
technical assistants made to the discovery?

E: My question is a difficult one and I know that, Dr. Waksman, but I would like you to be as precise as possible . . .
with the four or five names that you have mentioned.

w: Clara Wark, the technical assistant, it is difficult to say. Their
names do not even appear on the papers . . . It is very difficult to say they made one percent or 20 percent. It is very
difficult because they were concerned with several problems. They were not working exclusively on this one problem, you see, and therefore I could not answer that question.

E: Can you answer the question as to Miss Randolph?

w: No, as I said, she was one of the technical assistants.

E: But she was one of the 20 you mentioned?

w: No, the four or five.

E: I am talking about the four or the five [lab assistants] who
worked on streptomycin out of the twenty. [Eisenberg was
trying to get Waksman to repeat that Schatz's contribution had been "important."]

w: It is difficult to say Schatz made an important contribution.
If I would say my contribution would be roughly, let's

say, 75 percent, Schatz's contribution was perhaps
10 percent.

A few minutes later, Waksman changed his mind again. "As I said,
Schatz was the key figure because I depended upon him for many crucial
experiments . . . As I said, like testing the cultures against bacteria; like
production, when I was in a rush and in a hurry to produce material before the Merck people came into the picture, he worked day and night. To
be frank with you, I considered him to be one of my bright students"

Eisenberg moved on. He asked about Doris Jones and the parable of the
sick chicken. It was he, not Schatz, Waksman stated, who had received
the D-1 streptomycin-producing culture from Jones. And it was he, not
Schatz, who had selected the culture to be isolated.

> W: Dr. [Fred] Beaudette said to her, "Doris, here are colonies of
> organisms that Dr. Waksman has been looking for and has
> been interested in. You better take them to him." She brought
> the plate [petri dish] . . . I turned it over to Schatz and I said,
> "Now, will you please isolate these colonies and test them?"
> That was the beginning of the streptomycin story.

Schatz noted in the margin, "NOT TRUE."

> E: When did this conversation between Dr. Beaudette and
> Miss Doris Jones take place with reference to this plate to
> your knowledge?
> W: Roughly about the middle of August, 1943.
> E: When to your knowledge did Miss Jones bring that plate
> to you?
> W: Within a day or two.

Eisenberg made a note: "All this is untrue and Waksman knows it is
untrue because he was in Woods Hole in August and not around. Hence
everything is 'roughly.'"

> E: When did you turn that over to the plaintiff?
> W: Immediately.

E: You turned that plate over to him with certain instructions?

w: That's right.

E: Where were those instructions given?

w: Verbally.

E: Where?

w: In the laboratory, in the main laboratory upstairs.

Schatz wrote again, "Not True."

E: Was anyone present at the time besides you and the plaintiff?

w: I doubt it.

E: I asked you what strains produced streptomycin. You have been talking about a culture.

w: That's right, D-1, that was the beginning of all the streptomycin.

E: Who isolated that strain?

w: Isolated?

E: Yes.

w: Physically?

E: Yes.

w: Albert Schatz picked it up from the plate.

E: Who selected the strain to be isolated and tested?

w: I selected it and I gave him the instructions to go ahead.

"Not true," Schatz wrote.

E: Who tested it?

w: Albert Schatz, the first test; that is why, as I said, he was physically the agent that happened to do that, and that is why I recognized his contribution.

Eisenberg then picked up the money trail. Waksman said that at the time of the discovery of streptomycin he was being paid $10,000 a year by the university. In 1939, he had a verbal agreement with the Rutgers comptroller that whatever developed of a practical nature in his and Rutgers' deal with Merck, he would get a "certain small commission" for managing the affairs. It was a "loosely worded" agreement, nothing on paper. It was assumed that he would use the money for research work. He could put it "in my own pocket or give it for fellowships or assistants. It was up to me."

Eisenberg asked, "How much?" And Waksman hesitated.

> WATSON: Tell them . . .
> W: In 1949 . . . about $80,000, roughly.
> WATSON: Give the total for the whole thing.

Waksman then gave the total. It was about $350,000, he said. He said he paid $180,000 in taxes. That left him with $170,000.

CORNERED, WAKSMAN WENT on the offensive, a totally new attack that caught Eisenberg by surprise. Still under oath, he claimed, for the first time, that Schatz's 18-16 strain of *A. griseus* might not have been an original discovery, but rather a mere laboratory contamination from Jones's strain, D-1. He also suggested that Schatz had falsified his lab notes. The "evidence" was a missing page in Schatz's laboratory notebook.

> E: You have given us to understand that in August 1943 there was only one strain that produced streptomycin?
> W: That's right. 48 hours later—
> E: I want to know whether my statement was correct . . . When, to your knowledge, was the next strain found which produced something that later was called streptomycin?
> W: Two or three days later. Albert Schatz came up and brought another culture which was called 18-16. He said he isolated that culture from the soil. I assumed that was so, but it is quite possible that that culture could have come, because 48 to 72 hours is a long enough period of time. It could have come willfully or un-willfully from the first culture [which] could have contaminated the soil [samples] . . . As I

said, we gave credit in my report that 18-16 was another culture producing streptomycin. We have recently shown that the second culture could have easily been derived from the first culture.

E: When was that shown? You say "we have recently."

w: I mean I had one of my assistants purposely do that.

E: When, approximately?

w: Last spring.

E: You mean by that the spring of 1949?

w: Yes, I have personally, myself, taken the culture of *Streptomyces griseus* and transferred it, not being very careful about handling it, and then plating out the soil that was standing around the laboratory, and I picked up that culture from the soil; so it can be done.

Schatz wrote in the margin, "Nonsense. Never done."

THIS WAS AN entirely new allegation. Waksman had always concluded that the two strains were entirely separate, that there had been no contamination. In a joint article with Schatz in 1945, he had written, "The almost simultaneous isolation of the two cultures and the appreciable difference in the activity between the cultures at the time of isolation constitute evidence that they were independent isolations, rather than that one was obtained as a spore contaminant of the other."

In a collection of articles about the discovery titled *Streptomycin*, published in 1949 but actually finished in the fall of 1948, Waksman had been quite clear on the origins of the two strains. "Two cultures of an organism were isolated by Schatz and Waksman in September 1943 ... They were designated as 18-16 and D-1, were identical both in their ability to produce the same type of antibiotic substance and in their morphological characteristics. They were isolated in different rooms, in different buildings on the campus, and within two days of each other, thus excluding the possibility of one's originating from the other as a contaminant."

Now, in an effort to bolster the contaminant story, Waksman introduced another, more serious doubt about Schatz's experiments. During cross-examination toward the end of the deposition, Watson deliberately led Waksman into a story about a missing page from Schatz's lab notebook. Eisenberg made a note: "Funny business."

> WATSON: With respect to this [notebook] No 2, Dr. Waksman, in this book marked Exhibit P-8 for identification, one page is torn out. It is between the page marked 52 and 53, and the page marked 54 and 55. Will you please state what your knowledge is of that missing page, if any?
>
> W: Yes, that has a very interesting story behind it, Mr. Watson. Schatz, just before he left the laboratory, I began to get certain rumors that something is happening. Somebody is spreading rumors claiming a major portion of the credit for the discovery of streptomycin. One fine day, one of my assistants reported to me that a gentleman by the name of Martin, presumably a cousin or uncle of Schatz, broke into the laboratory and carried off Schatz's notebooks and kidnapped Schatz himself. I could not understand what was happening.

"Nonsense," Schatz wrote.

> W: 48 hours later, or thereabouts, Schatz appeared with the notebooks. I asked him, "What is going on here, anyway?" He said, "My family has been persecuting me. They have been after me that I should try to get more credit, that you are not giving me enough credit for my work."

"Not true, not true," wrote Schatz.

> W: I said, "Now, look here, have I not given you enough credit? What do you think? If you feel in any way that I have not given you enough credit, why didn't you come to me? You knew very well that I bent over backwards to give you all the possible credit." He said, "Yes, I know, but my family, I will have to leave them. I will have to run away from them.

I will have to run away because they are controlling me. They are forcing me to do things that I do not like to do."

"Nonsense," Schatz noted.

w: I said, "Where are the notebooks?" He said, "Here are the notebooks." I didn't pay any attention and of course I wouldn't go over every page to see whether it is there . . . Later, when I checked up in the notebook, I find that the crucial page—you asked me before about the second culture 18-16—the crucial page . . . between August 25th and September 10th there was a crucial experiment and that page has been very neatly taken out, and worse than that, Schatz with his own handwriting very carefully corrected all the corresponding pages . . . Well, then there is a question here for Sherlock Holmes. I began to go carefully through again and what do you think I discover? A very interesting psychological problem. The very date when the report is made about the chicken throat culture, I find a note on the very page where the first experiment is recorded, I find a note written by Mr. Schatz a year and a half later on February 2, 1945: "Miss Jones brought over the chicken throat and swabbed the plate while I was picking actinomycetes or after I have selected them for transfer. Therefore 18-16 and D-1 are separate isolations." Now tell me—

Watson interrupted, "Never mind the comments. Just give us the facts."

E: What was the name of the assistant who reported [the missing page] to you?
w: That I wouldn't be able to tell you, because at that time Schatz was no longer working in our main laboratory.
E: You don't recall who that was?
w: No.

E: Were you sufficiently familiar with that particular notebook to know that a page that had been taken out contained something on it when it was taken out?

W: Well, of course. I used those notebooks in writing up the papers. Those notebooks, as I said, Schatz was the force that wrote those lines, but I knew not only every word, but every idea behind it.

Eisenberg asked what the missing page might have contained.

W: That page may have contained information of critical importance to the isolation of culture 18-16 as related to culture D-1, because that is the page where it would logically have been described . . .

E: But you are not saying that it did contain such information?

W: No, of course I am not.

E: Was anyone present when you talked with the plaintiff about the return of the notebooks or the breaking in incident that you talk about beside you and the plaintiff?

W: I don't think so. As I said, he came to my office and cried like a baby. He cried and he said, "My God, Dr. Waksman, you have given me far more credit than I deserve. I have got to run away from my family." I said, "Well, that is for you and your family to fight out, but please don't drag me into it."

Schatz wrote, "Not True. Incredible!"

E: That is all.

The hearing adjourned at 2:50 P.M. It would be a vital part of Schatz's case.

UNDER OATH, WAKSMAN had reluctantly conceded that Schatz's work had been "important," then retracted that judgment, and then restored it. He had revealed, for the first time, the personal fortune of $350,000 that he had amassed from the royalties of streptomycin. He had reinforced the doubt about who had given Schatz the culture from the chicken's throat, the so-called D-1 strain, and he had then introduced two entirely new doubts: about whether the second strain, 18-16, might, in fact, have been the same as the first, merely a contamination from D-1, and about the missing page in Schatz's notebook. In the process he had accused Schatz's Uncle Joe of unlawful entry into his laboratory and of stealing a lab notebook, apparently with the intention of doctoring Schatz's experiments.

Schatz's notebook showed that a page had, indeed, been cut out. But there was no evidence of any break in the experiment that he had been conducting at the time—Experiment 11—in which the two separate strains of *A. griseus*, D-1 from Doris Jones and 18-16 from heavily manured soil, had produced streptomycin. The missing page was at the end of Experiment 11, but by then both D-1 and 18-16 had already been recorded separately as showing clear zones.

The page of Albert Schatz's laboratory notebook showing the start of Experiment Eleven. (Special Collections and University Archives, Rutgers University Libraries)

Watson would later point out to Waksman, if he didn't know it at the time, that the missing-page allegation was "insignificant."

And the story of the break-in by Uncle Joe and the theft of a key notebook could not have been true because Watson's brother, Dudley Watson, another Rutgers Foundation lawyer, would later reveal the notebook in question had been with the Merck lawyers on the date specified by Waksman. They had asked for it to complete the streptomycin patent application.

EISENBERG WOULD HAVE a chance to pursue Waksman's new allegations later, when he examined Doris Jones and the pathologist Dr. Fred Beaudette. For the moment, he was anxious to take advantage of what they had found out about the money.

In the tradition of the plaintiff's bar, juicy tidbits extracted in depositions are fair game. Eisenberg immediately leaked the story about Waksman's royalties. By this time, Waksman and his wife were in Europe, out of the public eye, but the story gave Schatz new confidence that he would win the case. His somber mood changed. He wrote to Jones,

> People, however, don't seem to realize that I would never have instituted litigation if I did not feel confident of proving my charges in court. The simple fact of the matter, Doris, is that Waksman is through!!
>
> He's in Europe now, probably doing a bit of politiking [sic] in an effort to collect a Nobel Prize. Very likely he is looking upon this as a last straw, with it and the attached prestige, he no doubt feels he has a chance. But it will not help him. His greed and lust for fame will convict him.
>
> You have no idea, Doris, as to the evidence on which our case is based. For example, I will tell you in the strictest confidence Selman A. Waksman has to date pocketed $350,000 of the total royalties derived from the streptomycin patent!!!
>
> Can you imagine what [people] will think when this information is made public!!! You see Waksman didn't give all the money away as he

publicly led the world to believe. Privately, he feathered his bed, and feathered it well . . .

I wanted you to know that I haven't gone off half-cocked on a willow-the-wisp [sic] chase. It is quite true that this litigation will result in the ruination of a professional reputation, but it is Waksman not I who will be ruined.

Schatz thanked Jones for offering to testify on his behalf but thought that it would probably not be necessary because his attorneys felt he had "an excellent chance of victory." In the event that Jones was needed, he told her, "we would want you only to tell the simple truth as you honestly recall it. That is all we want everyone to do.

"Unfortunately, though," he continued, "Selman A. Waksman is not cooperating in this respect. But his deposition is so full of contradiction and false statements (which we can refute with publications and even his own book) that he may well destroy himself on the witness stand. Waksman may very well prove to be our most valuable witness."

Jones replied that she had been "going around like the cat that swallowed the canary—just itching to tell my fellow 'scientists' the great news, but I was true to your confidence and kept the information locked up and gossip tite [sic]. It certainly is a confirmation of the opinions I personally held toward our friend—even though it is rather shocking to find he is not only a liar but a money conscious fool insofar as he couldn't even be decent about the dollars he grabbed unto himself. He had to make out to the world that he was a mild and beneficent man of great ideals and humble desires. Pooey!"

The only newspaper to fully understand the Rutgers PR legerdemain was the *Passaic Herald-News*. In an editorial, which may or may not have been influenced by Uncle Joe (there is no evidence either way), it pointed out to its readers some basic facts. The headline was RUTGERS IS TOO SMART FOR ITS OWN GOOD.

It does not reflect credit on Rutgers University to have allowed the people of New Jersey to get the impression that all royalties from the sale of streptomycin had been turned over to the university by Dr. Selman Waksman.

That, it now develops, was a false impression. Out of $2,600,000 in profits received by the Rutgers Research and Endowment Foundation, Dr. Waksman has been paid $350,000.

The payment of this sum to the distinguished scientist is, beyond doubt, defensible. He is probably entitled to every penny of it. But that isn't the point. The university allowed the public to assume he wasn't getting a nickel from the proceeds.

The admission that Dr. Waksman has shared in the profits was made at a closed pre-trial hearing in the suit brought against Dr. Waksman and the foundation by Dr. Albert Schatz, formerly of Passaic and now of Brooklyn. It was Dr. Schatz who, working under the guidance of Dr. Waksman, found the mold which produced streptomycin.

The admission throws a completely different light on Dr. Schatz's suit. He at first appeared to be a young man jealous of the honor that had come to Dr. Waksman, and properly so, as a result of the years of research that led to streptomycin's discovery. But so long as anyone is to profit personally from the discovery, he is justified in asking for a cut of the proceeds.

Rutgers must get over the idea it can hoodwink the public. The university's policymakers could read with profit what Abe Lincoln had to say about that.

AT RUTGERS, WAKSMAN'S deputy, Robert Starkey, kept his boss informed of the progress of the case while he was in Europe. He sent Waksman newspaper clippings, letting him know that his earnings from the patents were now public. The news, Starkey said, "was not presented in a favorable light—as you would expect—for they made no attempt to justify your receiving the funds or to indicate that you had made any other disposition of them other than for personal uses."

He was frank about the harm done. There had been "widespread consternation and the effect was not favorable." He said the general feeling was that "the public had been deceived into thinking that all of the royalty funds had been given to the university." The opinion had also been expressed that since this had been concealed, there might be "other information that was of general public interest but was not being mentioned."

"I am sure that your friends have no feeling that you are not justified in having any amount of money from the patents that you might wish," Starkey continued, "but they feel hurt that they have been misled with regard to disposition of the funds. A statement regarding the arrangement at the time of the announcement of the Institute of Microbiology would probably have been well received."

People had seen the reasonableness of the arrangement, "but there are probably a great number of people in important positions that do not understand what happened and are apt to consider the matter as reflecting upon the integrity of the persons involved. We have been in a period of unfavorable publicity which may introduce some difficulties in securing public support for the Institute of Microbiology."

ON JUNE 7, 1950, while Waksman was still in Europe, Eisenberg deposed Dr. Fred Beaudette, head of Poultry Pathology. Beaudette had accepted Doris Jones as a graduate student in July 1943, the same month that Schatz had started work on his Ph.D. in the basement. Her task was to look for antibiotics that acted on viruses. She was often in Schatz's lab, where he taught her the basic techniques—how to grow bacteria on a petri dish and isolate the promising ones. Beaudette showed her how to look for viruses in the gullets of chickens.

Eisenberg began by asking Beaudette if he had ever personally turned over to Waksman any of Jones's plates. He had not.

E: Did you ever suggest to Miss Jones that she do so?
B: I believe I did.
E: Do you remember when? The month?
B: No, sir.
E: Was it 1943?
B: I presume so.
E: Do you remember what plates you suggested that she turn over to him?
B: No, I can't say that I do, except that when she had a plate

on which there was a growth and there was some evidence
that it might have activity and ... I suppose it was suggested
that it might be turned over. The suggestion was not necessary. That would have been done automatically.

E: Were you present when it [the plate] was turned over to
Dr. Waksman?

B: No.

E: You never actually saw him receive physically any plate
from Miss Jones.

B: No, I did not.

On September 26, in Newark, New Jersey, Eisenberg deposed Doris
Jones. She said she had spent most of her time in Dr. Beaudette's lab
and had not reported "very frequently" to Dr. Waksman. Some of the plates
she used in her work she discarded, and some she gave to Schatz because
she knew what he was working on. She saw him once or twice a week,
and he asked her for any plates she no longer needed, and she gave him
"maybe fifty." These were plates she would usually have thrown into the
lab's trash.

E: Are you aware of the fact that one of the two active cultures of *A. griseus* was isolated from one of the several
plates you gave him?

J: Yes.

Jones added that Schatz had told her that he had obtained an active
culture from one of her plates and that the other one had come from a
field source. Russell Watson objected that this was hearsay and could not
be used in the trial, but Eisenberg pressed on. He asked her if she knew
who had first isolated the *griseus* strain that produced streptomycin.

J: Al Schatz.

E: Did you participate in the isolation?

J: No, other than the fact that I gave Al a plate, several plates.

Having established that Jones had given the plate directly to Schatz,
Eisenberg also wanted to ask about the conversation she had had with

Waksman in the spring of 1946 about the credit due to Schatz for the discovery.

Jones said that the conversation had taken place in Waksman's office and that Waksman had told her "confidentially" that the reason why he didn't let Al have more credit for the discovery was that he was so aggressive, and if he were allowed this credit, it would go to his head. "That was why he hadn't pushed Al's name.'"

Eisenberg now understood that Jones would be a key witness if the case ever came to trial. Watson had to find out whether she would corroborate, or at least not refute, Waksman's concocted story about the sick chicken.

Watson did not know that Jones had already made a statement to Uncle Joe's friend Max Bromberg about the affair. She had clearly stated that she had given the petri dish containing actinomycetes to Schatz. She had carefully passed it to him through the basement-lab window, and he had done the isolation with this dish as he had with so many before.

Now, Watson and Samuel Epstein went to Berkeley to interview Jones. They arrived at her house one evening in late April. Jones later told Schatz about it in a letter, reconstructing, as best she could, the conversation.

"Mr. Watson and Sam," she wrote, "proceeded to explain to me the facts of the case, read the complaint and the answer or what they told me were the important parts, and then Mr. Watson launched into a series of questions trying, as he said, to 'get at the true facts' because the only people who know the truth were the students who were working there. They wanted to know whether Waksman 'closely supervised' his students."

That depended on the student Jones had replied. "Those who had a head to think with and whom Dr. Waksman recognized as being capable were not very closely supervised; the student with ability did independent work. When I began there [with Dr. Beaudette], for instance, I hardly saw Dr. Waksman. Not to say that I had ability, or that I thought I had."

The conversation continued, she recalled, with Watson asking whether it was a long-term project and her replying, "I would say all students were working on a long-term project. Everyone knew, however, that the antibiotics were the center of interest."

WATSON: So you were hired to work on a specific problem?

JONES: Yes, but the details were not laid out in any strict order. [Jones noted to Schatz, "He was trying to get me to say I was a technician doing *only* things I was told to do."]

W: Did you report your results to Waksman?

J: At intervals, not from day to day, especially at first, since as I told you he wasn't there for a while.

W: How long was he away?

J: I don't remember exactly.

W: A week? 4 or 5 months? Can you judge?

J: I suppose a month, I think August. [She noted to Schatz, "He must have known the answer already anyhow."]

W: Whose idea was it to streak chickens' throats?

J: I believe—though I'm not positive—it came from Dr. Beaudette.

W: So, Dr. Beaudette directed you to streak them?

J: I can't remember. Possibly it came as an outgrowth of a conversation between both of us concerning the problem.

W: Whose chickens were they?

J: I certainly don't remember. They were perfectly normal chickens.

EPSTEIN: Were there actinomycetes on the plate [petri dish] you gave Al?

J: Yes, there must have been several. There was a mixture of organisms.

W: Who told you to give Al the plates?

J: I believe Al asked me for discarded plates.

W: Did Al ever discuss streptomycin with you?

J: Several times. I can't remember details offhand. I'd have to think longer.

W: Did he ever tell you Waksman threatened him?

J: He may have, I can't remember offhand. Too many years have elapsed and I'd have to sit down and reconstruct too many details to tell you now. [She noted to Schatz, "I didn't want to give any information."]

to Waksman would be "a pivotal issue" in the trial. Rutgers and Dr. Waksman faced a potential public relations disaster. It might be better for everyone to settle the suit, they said.

As Rutgers had no experience of such matters, Merck's John Connor, then vice president and head of the law department, was invited to present a proposal for a settlement. In discussions with Waksman and Watson, Connor had suggested that Waksman share the funds he had received, and future funds, not only with Schatz but also with "all others who had aided him in the research project," even the glassware washers in the laboratory.

In Connor's proposal, Schatz would now be only one recipient among twenty-four others, including almost everyone who had ever worked on antibiotics in Waksman's lab, and some who had never worked on streptomycin. The redistribution of the royalties took into account that Waksman, who had already received nearly $400,000, had volunteered to cut his share in half—to 10 percent. Starting from the fourth quarter of 1950, the proposal was to distribute most of the other 10 percent among fourteen of Waksman's researchers who had worked on antibiotics. Onetime bonuses, of up to $1,000, would be given to researchers who had been, in Waksman's view, less involved in the antibiotic program, like Doris Jones, but also included lab workers and Waksman's secretary.

In consultations among the lawyers the two sides proposed that Schatz should get the greatest share, after Waksman. He did, after all, risk his life. Russell Watson agreed "inasmuch as Dr. Schatz was one of the patent applicants, notwithstanding that in Dr. Waksman's opinion, Dr. Woodruff's contribution was more important."

In return, according to the proposed deal, Schatz would drop the charges of fraud and duress against Dr. Waksman and would assign the outstanding foreign patent rights for Canada, New Zealand, and now Japan, plus "other foreign patent rights, if any." For this he would be paid a lump sum of fifty thousand dollars.

Neither Schatz nor Eisenberg was privy to these initial consultations, so a draft was now sent to them for approval. Eisenberg pushed for more money for his client, of course, leaving the final amounts up to the judge.

The same group met again the day after Christmas. The trustees wanted to settle the case, but only if Waksman was "sincerely and enthusiastically" in favor of such a move. Waksman said he would prefer to "give

Schatz a kick in the pants" by going to trial, but he recognized that to do so would mean a great deal of "unpleasant publicity" for Rutgers, as well as for himself.

The lawyers for the two sides proceeded rapidly to settlement, a task that Watson undertook with "natural misgivings," and he issued a personal statement to the Rutgers Trustees. He still felt that "neither Waksman nor the University had committed any moral or legal wrong" in their dealings with Schatz. The "failure to inform the public" of Waksman's private deal with the foundation "was a mistake, but this was hindsight not related to the legal or moral issues of the Schatz case." He was "confident" that if the case had gone to trial, it would have been decided in favor of the defendants. Like Waksman, he yielded to others who feared the unfavorable publicity.

On December 30, the case was settled. Waksman's share was pegged at 10 percent. Since the sums would still be large, he arranged for half of that to be used to start another foundation—his own Foundation of Microbiology. As Hubert Lechevalier would put it, this new foundation would permit "him and Mrs. Waksman, with the assistance of three other trustees, to play the role of Maecenas." Schatz was accorded the legal status of "co-discoverer" and given 3 percent of the royalties, or about twelve thousand dollars a year—more than twice his salary at Brooklyn College. He was also awarded $125,000 for assigning the foreign patents to the Rutgers Foundation. His lawyers took 40 percent. The other twenty-four past and present workers on Waksman's staff, many of whom had never worked directly on streptomycin, were awarded sums ranging from five hundred dollars to a smaller share of the royalties worth a few thousand dollars a year for the life of the patent, or fifteen years.

In a statement, Judge Thomas Schettino, himself a Rutgers alum, said he was "in the firm belief" that the settlement was "an excellent one" because it had been made in "a spirit of appreciation of the adversary's feelings and reputation." He praised the attorneys for confronting the problem "in a spirit of cooperation and good will." He had urged the attorneys separately to settle "if it could be done with grace and with credit to all parties concerned."

Striving to be evenhanded with the two newly affirmed "co-discoverers," he said that Waksman had "on many occasions proven himself a great humanitarian and a great scientist. The terms of this settlement add additional prestige to his reputation." In assessing Schatz, he deferred to the two principal lawyers. They had told him that Schatz had a brilliant record. "They advise me he will go far in his field. We wish him well."

The judge beat the drum for his old college. "The Trustees of Rutgers University have shown once again how high is their caliber. I am proud to be, like Dr. Schatz and Dr. Waksman, an alumnus of this great State University. It has given generously of its funds to assist in the settlement. Such an act is worthy of its past record." Rutgers' president, Robert Clothier, said that it had "never been disputed that Dr. Schatz was a co-discoverer of streptomycin." That had been "a matter of public record since 1945, when Dr. Waksman and Dr. Schatz applied jointly for the streptomycin patent."

Then Clothier gave the official version of the patent's history. Dr. Waksman's hopes of preventing a monopoly by assigning the patent to the Rutgers Foundation had been fully realized, he said. "In 1946 streptomycin cost more than $25 per gram; today it wholesales for less than 35 cents." As streptomycin proved itself in 1947 and 1948, the royalties began to accumulate "in unanticipated volume." This prompted Dr. Waksman to request the Foundation to reduce his participation from 20 to 10 percent.

THUS ENDED ONE of the most contentious legal battles ever waged between a professor and his student. It was a stunning victory for Schatz and his colleagues. There seems little doubt that if he had not stood up to Waksman, then Waksman would never have felt it necessary to share his spoils with any of them.

On December 30, the *Newark Star-Ledger* editorial was headlined "He Finally Gets Credit." It began, "It is heartwarming to learn that a 30-year-old Brooklyn College professor will share credit with Dr. Selman A. Waksman of Rutgers University for discovering streptomycin." But, the paper added, "it will take more than long-winded explanations to convince the public that Schatz didn't suffer a grave injustice as a reward for his part in this important advance in medicine. A change in the system of crediting scientific discoveries so that the rewards are fairly distributed appears to be in order." The *New York Times* quoted a statement by Schatz expressing his gratitude at being "recognized as the co-discoverer of streptomycin." In agreeing to the settlement, he had been "influenced by the fact that the [Rutgers] Foundation and the public trust imposed upon it might be

Albert Schatz and his lawyer, Jerome Eisenberg, after their victory.
(Courtesy of the Newark Public Library)

jeopardized if the litigation were to continue to its ultimate conclusion."
The *Newark Sunday News* headline for December 31, 1950, was DR. SCHATZ
IS MODEST IN VICTORY. The young man had sued "to win formal recognition as co-discoverer of streptomycin. He was a slim, attractive, boyish
young man . . . with a winning smile, a keen sense of humor and a ready
tongue which delights in calling a spade at least a shovel."

Russell Watson, who was prone to taking a lawyer's gloomy view of
things generally, was especially upset about the *Newark Star-Ledger*'s comment that it would take more than Rutgers's "long-winded explanations to
convince the public that Schatz didn't suffer a grave injustice." He would
tell Waksman that, in his judgment, the newspaper accounts of the settlement generally "were grievously injurious to the University and to you, in
less degree to the University than to you. How long this public impression
will persist is uncertain."

Doris Jones wrote immediately when she heard the news. "Needless to say I was extremely happy for you and Viv. It must be a wonderful relief from all the tension you were living under and besides who can sneeze at the monetary comfort. Achou! Rutgers must indeed be proud of you."

Jones had saved a few clippings from the local papers. The general opinion was that Waksman "was a shrewd codger and you cleared away some of his manure pile to expose the true contents of the man."

The *San Francisco Examiner* had called to interview her about receiving her five hundred dollars. She had been so astonished, she had told the paper it must have the story wrong. But since then she had received a check. Jones saw the payoffs as a "clear maneuver to present a more generous picture and to protect Waksman from further litigation." The check and Clothier's statement "confirmed—or seemed to—my suspicions since Waksman expressed his 'long-felt desire' to share the proceeds with the individuals working on streptomycin and the antibiotics in general."

The "subtle part" for her, she said, was that the five hundred dollars was said to be a "flat payment" for her services for all work done up to October 1948. "I thus am well hushed-up should any other discovery become of financial value. Of course, I evaluate what I did at 2 cents and thus worry not a bit, but it certainly appears a stinking move—and especially so as Waksman presents the act as one of his own generosity, and Clothier bitches about how the suit held up the Inst. of Microbiology and lauds our fair boy [Waksman]. I guess you have seen the letters."

In those letters to all bonus recipients, Waksman, as he put it, had distributed the money to three categories of workers: "1. Those who have played a major, significant and direct part in the isolation of streptomycin, its evaluation in vitro and in vivo, and in its utilization. 2. Those who have contributed, through their scientific work, to the development of the antibiotic program as a whole which had a bearing upon the isolation of streptomycin. 3. Those who have assisted faithfully in carrying out this program." Jones qualified in all three categories. She had given Schatz the *griseus* culture from which he had isolated the D-1 strain. She had tested streptomycin in chick embryos. And she had assisted faithfully in the program. But Waksman had only acknowledged her "faithful assistance," and she had received the same as the glassware washers.

———————

MOST OF THOSE given awards accepted the money, even though several stressed they had not been directly involved in the discovery. Robert Starkey wrote Waksman that he was "very grateful for your consideration, and I am not unmindful of the fact that my direct contributions to the success of the project leading to the production of streptomycin were negligible." Boyd Woodruff considered his award to be the "equivalent of a gift from you, since I did not work directly on streptomycin and certainly have no direct claim to the royalties." One of the graduates, Dale Harris, thanked Waksman for "some of the best years of my life . . . your scientific curiosity, enthusiasm and integrity." Others, like Jones, heard about their awards through their local newspaper.

Corwin Hinshaw of the Mayo Clinic was "utterly surprised, even amazed, that my name had entered into the settlement . . . I would never in any way have pressed such a claim. I also have such deep respect for your scientific standing and for your thoroughly ethical conduct . . . Several of my friends . . . have shared my admiration of your willingness to forego [sic] the opportunity of becoming a very wealthy man in order to serve the interests of humanity. This adds significantly to your already established stature as a scientist and as a philanthropist in the truest sense of the word."

Hinshaw's partner, William Feldman, was one of two to refuse the money. He declined to accept "any royalties from the sale of streptomycin or any other drug which I have studied experimentally." His decision was "entirely personal and unequivocal. It has no relation whatsoever to the controversy which led to the court decree nor to any other persons involved in or affected by that decree." He said his work on streptomycin had been "a genuine privilege" and had continued "to be a source of deep satisfaction to me personally." He had used many substances from individuals and "pharmaceutical concerns" over the years of looking at TB, and they had been "supplied to us without charge and with complete freedom to work with and to report upon without restriction."

Feldman said that to have accepted would have been "in conflict with my concept of the principles governing my role as an independent investigator. For me to assign funds from such sources to charitable, educational or research institutions it would first be necessary for me to accept the funds . . . As one who has served science with devotion and much distinction for most of a busy and productive life you will, I am sure, appreciate my point of view and accept my decision with understanding."

During his twenty-three years as a member of the staff of the Mayo Clinic, he added, "I have been afforded complete freedom to conduct my research in the best traditions of an academic community. This is considered to be the sine qua non of my scientific endeavors."

Waksman wrote back, "I regret sincerely that you found it advisable to refuse the small royalty which the RREF offered, at my recommendation, to place at your disposal. This was *not* a court decree but a gesture of appreciation on our part." He had included Feldman in recognition of "the important part that you and Dr. Hinshaw have played in establishing the role of streptomycin in tuberculosis." But, out of his wishes, he had taken Feldman's name off the list.

Waksman was clearly troubled by a second rejection from a chemist named Walton Geiger, who had worked in Waksman's Rutgers lab. Geiger had been a graduate researcher at the time of the streptomycin discovery. His decision to reject the offer was written up in the *New York Herald Tribune.* He believed that proceeds from royalties should be plowed back into research, and in a letter to Waksman he quoted several academic references to support his decision.

Waksman replied to Geiger, "I am sorry that you felt compelled to do that. I have read carefully the papers that you submitted and could not find any justification for your action. However, I respect your wishes." He had wanted to argue with Geiger's decision, but Watson suggested it was now wise "for all of us to say and write as little about the Schatz case as possible in the hope that discussion will subside."

Part IV · The Prize

19 · The Road to Stockholm

ONE OF THE MOST SUCCESSFUL OF Rutgers' PR myths was that Waksman "gave up millions" in royalties from streptomycin; another was that he was almost fired at the beginning of World War Two. Feature writers from the popular weekly newsmagazines found such stories irresistible. Rutgers and Waksman worked hard to keep the myths alive, and at the beginning of 1952, when *The American Magazine* called on Waksman for a profile, the professor was in dazzling form, showing no hint of his bruising legal battle. The *American* had started life as a muckraker's journal, but by the 1940s it focused on social issues and human interest stories, with a star list of contributors, including Agatha Christie, Graham Greene, Dashiell Hammett, Upton Sinclair, P. G. Wodehouse, and H. G. Wells.

The reporter spent several days on campus, where he heard the story of Waksman's almost being fired in the winter of 1941–42. According to the story, Rutgers was short of funds and had to cut the budget drastically. Waksman's salary was a big one—$4,620. One penny-pincher suggested that was an easy cut. The fifty-three-year-old scientist was, in his opinion, just "fooling around with microbes." But Waksman was supposedly saved from the ax. Waksman knew that this tale was "about the biggest hoax that could have been perpetrated," but instead of correcting the story, he embellished it.

Asked why he didn't want to become a millionaire, Waksman replied, "What would I do with all that money? I'm too busy to be a millionaire. I have work to do." When the reporter moved on to his work on antibiotics, Waksman's "features lit up as they never did when we talked about money."

"Do you realize we haven't even made a beginning?" Waksman declared.

The reporter found Waksman "so completely absent-minded that the role might seem overdrawn if presented on the stage." The five-page feature, with a picture of Waksman in a white lab coat handling a test tube of microbes, wrote its own headline and subhead: "He Turned His Back on a Million Dollars: An Intimate Glimpse of a Distinguished Scientist Who Passed Up a Sure Fortune for the Greater Reward of Freeing Mankind from Disease."

The reporter credited Waksman, and Waksman alone, with the discovery of streptomycin. Albert Schatz was never mentioned. A single paragraph dealt with the lawsuit as a bid for money from a former student. Like other journalists before him, the reporter was easily charmed by the "small, stocky compact man . . . in rumpled and un-pressed clothes [and] high-buttoned, un-shined shoes . . . [who] badly needed a haircut." The story was Waksman's carefully crafted fable, half true at best.

OUT OF THE popular media spotlight, thousands of miles away from fawning American reporters, Dr. Waksman's career was under scrutiny of a very different kind. Behind closed doors, the faculty of Stockholm's Royal Caroline Institute was deciding who should be awarded the highest honor in the medical world, the Nobel Prize in Physiology or Medicine.

Each year since 1900, the institute has sent out requests for nominations from academics and practicing doctors. The nominations have to be submitted by February. They are evaluated through the spring and summer by a committee, then voted on by the twenty-five full professors of the institute. The award is announced in October, the month of Alfred Nobel's birth, and is presented by the Swedish monarch in Stockholm in the second week of December to commemorate the anniversary of Nobel's death on the tenth of that month.

In 1952, for the seventh year in a row, Waksman had been nominated for his work on antibiotics. The first nomination had come in 1946 from a professor of pharmacology at the University of California, San Francisco. Those were early days for streptomycin, which had only just been approved for general sale. The recommendation was for Waksman's "work on soil organisms, and the development of streptothricin and streptomycin." But he didn't even make the short list for at least one good reason. The year before, Alexander Fleming, Howard Florey, and Ernst Chain had been given the prize for penicillin. The Caroline Institute had already recognized the discovery of how antagonistic microbes can provide wonder drugs, and the second discovery, streptomycin, although full of promise for ending the scourge of tuberculosis, had not quite proved itself.

As the wonders of streptomycin became more established with each passing year, though, the pressure was building to notice its place in the galaxy of inventions. The next year, 1947, Waksman had been nominated five times, twice on his own for "work on antibiotics" and three times with William Feldman and Corwin Hinshaw. Those nominations had combined the two stages of the discovery: the "identification" of streptomycin and the "experimental and clinical investigation of its properties."

The word "discovery" only started to appear in the 1948 nominations. Of the five in that year, two were proposed by Turkish physicians from Istanbul, one by an Italian neuropsychiatrist from Rome, one by a German professor of surgery from Göttingen, and only one by an American, a professor of medicine from New York. The Nobel Committee for Physiology or Medicine determined that the discovery of streptomycin, which it attributed to Waksman alone, was worthy of a prize, but it was concerned about the real effectiveness of the drug and decided to wait for more clinical tests.

In 1949, and again in 1950 and 1951, the committee concluded that the clinical results were sufficient for a prize, but Waksman lost out to two brain researchers in 1949, to three researchers on adrenaline in 1950, and in 1951 to a researcher who had discovered a breakthrough in the treatment of yellow fever.

At the beginning of 1952, Waksman was nominated again—four separate times—and so, for the first time, was Albert Schatz. A Yugoslav professor of medicine, Jevrem Nedelkovitch, who had experienced the powers of streptomycin at his hospital in Belgrade, had read the scientific papers announcing the discovery of streptomycin and nominated Schatz, Betty Bugie, and Waksman, in the order that they appeared on the 1944 paper announcing the discovery.

During the summer, the Nobel Committee asked for opinions from two leading members of the Caroline Institute, Professor J. O. Strombeck, a prominent plastic surgeon, and Einar Hammersten, the institute's professor of chemistry.

Strombeck was asked to give his opinion on Waksman, who had been proposed in a general way for his work on streptomycin. On the basis of the success of the clinical trials of streptomycin in Britain, the United States, and Sweden, Strombeck concluded that streptomycin's effect, primarily against TB infection in humans, deserved a prize. In short, Waksman should be given the award.

Hammersten had a more difficult task. Because of his expertise in chemistry, he was asked to look at all the nominations: the set that proposed Waksman alone; the set proposing Waksman and two chemists, Karl Folkers from Merck and Oscar Wintersteiner, of New York; the set proposing Waksman, Feldman, and Hinshaw; and, finally, the nominations of Schatz, Waksman, Bugie, and Boyd Woodruff (for his work on the extraction techniques for streptothricin).

Hammersten then reviewed the evaluations of Waksman going back to the "preliminary assessment" in 1946, when Waksman's entry had not made the crucial transition to a "special assessment" by the Nobel Committee (the short list). In 1947, Waksman, Feldman, and Hinshaw were nominated for the "identification of streptomycin and experimental and clinical investigation of its properties and of other antibiotic agents" together by three professors of medicine at the Mayo Foundation Graduate School of the University of Minnesota. Hinshaw was invited by the Nobel committee to go to Stockholm to show his data. The three men qualified for a "special assessment," and even though streptomycin was now available to the public in America, they were not found worthy of a prize because the committee still wanted to see more data of streptomycin's effects on human infectious diseases.

In 1948 and through 1951, there had been a clear consensus that Waksman alone was worthy of the prize, and Hammersten started by grading Waksman against the chemists.

For this, Hammersten went back to the discovery of streptothricin, the second antibiotic found by Waksman and Woodruff in 1942. The method they had used to extract and purify the drug had consisted of adsorption onto carbon and the elution, or removal, of the antibiotic with weak acid. This had not been a groundbreaking method, as Woodruff himself had made clear in his deposition during the lawsuit. But under examination by Eisenberg, Woodruff had gone to great lengths to explain that the novelty had not been in any one method, but in the "association of steps" taken.

Eisenberg had asked, "Were the methods that you used, or were used in the isolation of the antagonists that produced actinomycin and streptothricin, novel?"

"Well, novel is a hard word to define in laboratory work . . . ," Woodruff had responded.

Later, Eisenberg had pressed the question, and Woodruff had replied that it was "the association of steps, known steps in the right order leading to the isolation of a concentrated purified or crystalline material which can be considered novel."

> EISENBERG: In other words the steps by themselves, taking
> each step separately, was not a novel procedure?
> WOODRUFF: That's right.
> EISENBERG: But the order in which these steps were taken you
> say was the novelty?
> WOODRUFF: Yes.

But Hammersten, under the Nobel protocol set up for evaluating candidates, was required to consider "only scientific publications concerning the work of and by prospective candidates."

He relied solely on the published scientific papers, and not at all on the lawsuit transcripts, in making his evaluation. He argued that the chemists had done complimentary work and would have to share a prize, but also that their isolation and purification work had not gone far enough beyond Waksman's original crude attempts at purification to warrant an award. What the chemists had done was extraordinary work, but was not, in the end, prize-worthy, because Waksman had shown the way. Hammersten concluded that credit should go to Waksman for using a method of extraction for streptothricin that could be later applied to streptomycin.

Hammersten then turned to the question of the part played by Waksman's collaborators in these extraction experiments. In this early work, he noted, Woodruff was a *"medarbetare"*—literally "with work," and with a general meaning in Swedish of "colleague" or "collaborator," but when applied in this sense, of a master and his apprentice in the laboratory, definitely meaning "apprentice." The signal to the Nobel Committee of the Caroline

Institute was clear, and Hammersten did not have to spell it out: Wood-ruff was too junior in rank to be considered as an equal to Waksman, and therefore the credit should go to Waksman alone. If Hammersten had considered him as Waksman's equal, he would have written *"collega,"* a colleague of equal merit.

So, thus far, Waksman alone had been approved for a prize, by Strom-beck for the discovery and by Hammersten for his part in the extraction. That left yet another assessment of the relative parts played by Waksman, Bugie, and Schatz.

Schatz was the senior author on both the 1944 papers, but, as is the custom in such publications, the text gave no clue as to who had done what in the experiments leading to the discovery. This was, as always, a complex matter of personalities and actual lab work. The experiments had been written up impersonally in the passive tense. The paper said that streptomycin "was isolated from two strains of an actinomycete." The person who had isolated the strains could be identified only by reference to the laboratory notebooks. But, again according to the rules, Hammersten could not delve deeper than the scientific papers—he did not look at the lawsuit file, the patent application, or the lab notebooks.

In examining the first paper, announcing the discovery of streptomy-cin, Hammersten therefore missed the later affidavit that Betty Bugie had given to the Patent Office acknowledging that she had had "nothing to do with the discovery of streptomycin." He didn't consult Schatz's Ph.D. thesis, or Schatz's laboratory notebooks, which showed, beginning with Experiment 11, that Schatz had been the first to isolate streptomycin and had also prepared crude extracts by the traditional method of adsorption on carbon and then elution with acid.

From the second paper, reporting streptomycin's effect on the TB germ, on which Schatz was the first and Waksman the second author, Hammersten could not have known that it had been Schatz, working alone in the basement laboratory, who had risked catching TB while testing streptomycin against the virulent strain of the disease.

Finally, Hammersten reviewed Waksman's work on the *griseus* strain that Schatz had used to make the discovery. Hammersten concluded that Waksman alone had isolated *S. griseus* in 1919 and in 1942 had found, together with other students including Woodruff but not Schatz, that the actinomycetes were especially promising producers of antibiotic substances.

This paragraph not only has the year of Waksman's isolation wrong (it was 1916), but is a reductio ad absurdum view of the history of the *griseus* strain. As mentioned in earlier chapters of this book, Waksman was not the first to isolate *griseus*. It was first isolated, and named because of its grayish color, by the Russian researcher Alexander Krainsky in 1914. Streptomycin was indeed isolated by Albert Schatz from a strain of *S. griseus*, but not the strain that Waksman had isolated in 1916, which did not produce an antibiotic. Schatz had found a new streptomycin-producing strain. In addition, Russian researchers had in fact already mentioned that the actinomycetes were promising producers of antibiotics substances.

Hammersten, bearing in mind the award of the Nobel Prize to Fleming, Florey, and Chain in 1945, also pointed out that Waksman had developed streptomycin further than Fleming had developed penicillin, and Waksman had been involved in the methods used to extract and purify streptomycin.

Hammersten summed up his review by saying that because of Waksman's well-known and leading position in the discovery of streptomycin, and because only Waksman's name appeared on the three most important papers announcing the discovery, he alone should be considered as the discoverer. Albert Schatz was a *medarbetare*, an assistant of inferior rank.

20 · "A Dog Yapping at the Heels of a Great World Figure"

IN THE WEEK OF OCTOBER 20, 1952, reporters from Swedish newspapers began calling Waksman for interviews. They had heard he was going to share the Nobel Prize in Physiology or Medicine. Then photographers called to arrange to take pictures. And on Thursday, October 23, at four o'clock in the afternoon, the Royal Caroline Institute announced that Selman Waksman, alone, had won the prize "for the discovery of streptomycin, the first antibiotic which is effective in cases of tuberculosis." The citation was specific. It was for *the discovery* itself. The next day, October 24, the *New York Times* ran a front-page story, WAKSMAN WINS NOBEL PRIZE FOR STREPTOMYCIN DISCOVERY. Waksman's prize "crowned decades of relentless effort that root back almost to the moment he set foot in America."

Reporters and television crews turned up on the cramped third floor of the Administration Building, where the sixty-four-year-old Waksman told them, "This is the culminating point of my life's work begun in 1915 with the study of a humble group of soil microorganisms, the actinomycetes. I feel particularly proud for the field of science which I represent, microbiology, and for the institution where I have done my major work, the College of Agriculture and Experiment Station of Rutgers, the State University of New Jersey." For the occasion, he selected his oft-repeated quote from Ecclesiastes about medicines coming from the soil. "The Lord created medicines out of the earth, and he that is wise shall not abhor them."

Waksman would remember the day as being filled with telegrams from all over the world, from former students, colleagues, well-wishers, tuberculosis sufferers and "foreign groups proud of the accomplishments of an 'immigrant boy.'" His office was "bedlam . . . swamped with reporters, photographers, radio and television groups." He "had to answer all sorts of questions." But his life was now so full of awards and honors that the award was really only "another surprise" in his busy life.

A few days later the Associated Press wire service chose Waksman as man of the year for science in its annual awards. President Dwight Eisenhower won for politics, Queen Elizabeth II was woman of the year, Ernest Hemingway was honored for literature, Marilyn Monroe for entertainment, and the boxer Rocky Marciano for sports.

In his interviews celebrating the award, Waksman did not mention Albert Schatz. But the newspapers did. The *New York Times* noted that Waksman was the fourth researcher in antibiotics, after Fleming, Florey, and Chain in 1945, to receive the prize. In a paragraph inside brackets, the *Times* also reported that while Waksman was widely credited with the discovery, he had acknowledged in 1950 that Dr. Albert Schatz, an assistant to Dr. Waksman, at the time was "entitled to credit legally and scientifically as co-discoverer."

The *Philadelphia Inquirer* went further. "Progress today," it editorialized, "in any field, but particularly in medicine, is usually achieved only by the co-operative effort of many persons." Much depended "on the individual researcher" who "may be the head of a project or one of the rank and file. If the latter, he deserves a share in the honor."

Albert Schatz had left his job at Brooklyn College and moved to a new research post at a small school, the National Agricultural College, in Doylestown, Pennsylvania. Fifty miles from Rutgers, in the converted farmhouse he now used as a laboratory, Schatz was stunned. He could not believe what the Nobel Committee for Physiology or Medicine had done. How could it have ignored the published evidence of his involvement— the two crucial scientific papers that named him as the senior investigator in the discovery, and the court settlement, on which the ink was barely dry, honoring him as the co-discoverer of streptomycin? He kept staring at the citation: *for the discovery of streptomycin.* He felt as though he had been dealt a blow that denied his "entity as a human being." At an Ivy League college, missing the award might have been taken in stride—there would be other opportunities to win a Nobel—but the leaders of the tiny

National Agricultural College decided to protest what they saw as a great injustice. Within a week, the college's vice president, Elmer Reinthaler, had drafted a letter to the secretary of the Nobel Committee for Physiology or Medicine, Professor Göran Liljestrand. Reinthaler expressed the "amazement" of the college's administration and faculty that the award had been made solely to Waksman.

Choosing his words carefully and respectfully, Reinthaler wrote, "We are certain that so distinguished a body as the Council of the Caroline Institute could not have been aware of, and yet ignore, certain most pertinent facts regarding the discovery of streptomycin and the original co-discoverers thereof." He cited the 1945 Nobel Prize in Medicine, which had been awarded "in the most equitable manner" to Fleming, Chain, and Florey. If penicillin could have more than one prizewinner, why not streptomycin? He enclosed a short history of the discovery, with the relevant scientific papers showing the part played by Schatz, adding that the college was "convinced that further consideration by your Council would be well warranted."

At the same time, Schatz launched his own appeal. He drew up a list of colleagues he thought might support him and prepared a "resolution" that he planned to ask them to sign. The resolution "assumed" that the nominations submitted to the committee had "neglected to stress the role played by Dr. Schatz in this discovery, thus depriving him of his rightful share in the prize." Addressing the matter of rank, in case the committee had given the prize solely to Waksman because he was the head of the laboratory, Schatz argued that he should not have been excluded from the award simply because his thesis had been supervised by Waksman. There was "ample precedent" in the history of Nobel awards to the contrary: the Polish-French Marie Curie, for physics, in 1903 (she had shared the award with Pierre Curie); the Swede Svante Arrhenius, for chemistry, in 1903; and the French Louis de Broglie, for physics, in 1929. The work for which they had received Nobel Prizes had been "largely embodied in their doctoral dissertations." So it was with Schatz's thesis on streptomycin.

"Under the circumstances and on the basis of the available evidence we must regretfully state that, in our opinion, an injustice has been inflicted on Albert Schatz," the resolution declared. The document quoted a passage from the 1950 book *Nobel, the Man and His Prizes*, which had been edited by the Nobel Foundation. Committee secretary Liljestrand had written in the book about an underlying intention of the Nobel to help promising young researchers by providing "such complete economic independence for those who by their previous work had given promise of future achievement that they could ever afterwards devote themselves entirely to research. While an award, therefore, to an old scientist at the end of a fruitful career would seem to be a well-deserved tribute to truly important achievements, it would scarcely harmonize with Nobel's own ideas." Kurt Stern, a professor at the Polytechnic Institute of Brooklyn, agreed to organize the petition.

The replies to the appeal came back swiftly but were not encouraging. Most scientists did not want to be drawn into the dispute, while some expressed outright hostility to the idea of an appeal, even opposing Schatz personally. They sided with the professor and chastised the student for being uppity. In many ways they were defending the system that had rewarded them. They did not want it undermined from below.

In an especially blunt letter, Albert Sabin, professor of pediatrics at Cincinnati's Children's Hospital, said that Schatz was behaving "like an ungrateful, spoiled, immature child." And when he grew up, he would regret what he had done. Sabin, a Russian immigrant who would become famous for the development of an oral vaccine against polio, wrote that, in his opinion, "Dr. Schatz should have considered himself to be an unusually fortunate graduate student in having been permitted by Dr. Waksman to participate in the great work he was doing. Any other graduate who might have been in Dr. Schatz's position with the inspiration and the tools supplied by Dr. Waksman would have achieved the same. The Nobel prize is not awarded for accidents. It is awarded for the discovery of important new principles which open up new fields of research. This Dr. Waksman has achieved. In this achievement, Dr. Schatz made no contribution."

A British chemistry professor, Maurice Stacey, of the University of Birmingham, echoed what many of his colleagues felt about rank. Schatz, while playing an important role, had been, after all, only a student. "I cannot agree that such a junior person should share the great honor of the Nobel Prize," Stacey wrote. "Surely this was given to Dr. Waksman in appreciation of a lifetime's outstanding work in microbiology. Whatever Dr. Waksman's personal faults, and they may well be many, scientists recognize him as one of the world's leading authorities in his subject and doubtless the Swedes had this in mind when they awarded him the prize.

If Dr. Schatz is as good as you appear to think he is, then surely his day will come in the future!"

The most disappointing response for Schatz came from his former professor C. B. van Niel, at the Hopkins Marine Station in California. Van Niel concluded that the discovery of streptomycin had been, like that of penicillin, a "happy accident" and, as such, was not worthy in itself of a Nobel Prize. He found "ample justification" for the selection of Waksman for the award based on his "many and major additional contributions." This "enormous output" was the work "not merely of a hard-working enthusiast and compiler, but of a true scientist, scholar, and master." In conclusion, he was "satisfied that this award does not constitute a serious injustice to Dr. Schatz." Van Niel sent a copy to Schatz with a penned note: "Dear Al, This is my well-considered opinion, not lightly arrived at. Very best to you, Vivian and etc. Yours, Kees."

One of the more thoughtful replies, and certainly one based on greater knowledge of the discovery than that of most of the other scientists, came from an unexpected source: William Feldman of the Mayo Clinic. He had not been on Schatz's appeal list, but had read a copy of the appeal sent to a friend. He sent two replies, the first to Schatz. He was "disappointed" that Schatz had not been named as a "co-recipient," he wrote. "It always seemed to me, without being in possession of the details, that your contribution to the discovery and development of certain important basic facts was quite indivisible from the contribution of Doctor Waksman."

In a second, formal reply to Stern, the organizer of the appeal, Feldman said that "from my knowledge" of Schatz's part in the discovery, "it would seem just and proper that any award recognizing solely this contribution should be conferred jointly" on Waksman and Schatz. But he asked whether it was "definitely known" whether the Nobel Prize was "especially in recognition of the *discovery* [emphasis added] of streptomycin, or was it in recognition of Dr. Waksman's distinguished achievements of a long professional career pertaining largely to investigations [into] the microbiology of the soil from which streptomycin finally emerged? This distinction is extremely important and the exact wording of the citation should be known before any protest can properly be formulated."

Overall, Feldman felt that the protest was "most unwise." The people in Stockholm should be given a chance to clarify the award first, he argued.

He declined to be a part of the proposed resolution, but he had identified the problem. The citation was for the "discovery of streptomycin."

AT SCHATZ'S COLLEGE, Elmer Reinthaler pressed the Nobel Committee for a reply, again listing the two key 1944 discovery papers plus a third paper from 1946 on the different strains of *S. griseus* that produced strepto-mycin, also with Schatz as the lead author. Referring to the committee's apparent lack of access to all the relevant material, Reinthaler wrote, "We think you will agree with us that mere procedural technicalities should not permit what would constitute a serious injustice to a scientist, and conceal from the scientific and lay world the facts surrounding the discovery of streptomycin."

On November 14, the committee replied. Vice President Reinthaler's letter had been discussed at a faculty meeting, and "it was generally regretted that part of the information given in your letter had not been accessible to the members of the faculty, since it had not been published in any scientific journal." The committee was apparently specifically referring to Schatz's doctoral thesis, which had laid out his work and, indeed, had not been published by Rutgers, but also to the patent and the settlement of the lawsuit between Schatz and Waksman. "It may interest you to know," the letter continued, "that numerous American colleagues who have been invited to make proposals about the Nobel prize have suggested the name of Dr. Waksman, though none of them has proposed Doctor Schatz." The committee told Reinthaler that the Nobel rules made it impossible to reconsider the prize. According to those rules, the decision made in October could not be altered and "no protest shall lie against the award of an adjudicating body."

ALTHOUGH WAKSMAN MADE no public comment about Schatz's petition, he was well aware of it. Several colleagues who had received copies of the letter to the Nobel Committee had written him expressing their support. One former student said that the appeal to the committee had "filled [him] with disgust," and he could "find no better place to file such a letter than with the trash in my wastebasket. Undoubtedly, such low attempts by little men must grieve you deeply, and it is entirely beyond my comprehension how a pupil who like a baby was taught the first steps in the field of micro-biology by a great Master, can consent to such blows ... The Nobel committee could have made no nobler choice and it was fully deserved."

Looking over the information that Reinthaler had sent the committee, Waksman was outraged. He told Russell Watson that he wanted to sue Reinthaler and Schatz for libel. Watson advised strongly against any litigation "of any kind at the present time, assuming that the quoted sections of the National Agricultural College letter are libelous." The college's appeal appeared "to have been abortive," and any "litigation would revive public discussion of the Schatz case with some degree of unpleasant notoriety."

Waksman was clearly agitated, imagining that somehow Albert Schatz was going to derail his Nobel Prize. But he had no cause for concern; a compromise was in the wind. A colleague, Stuart Mudd, a professor of microbiology at the University of Pennsylvania, had suggested a way out of the mess the Nobel Committee had created: Why award the prize specifically for the discovery? He had written the committee that he was "delighted" the prize had gone to Waksman and "could hardly imagine a more appropriate award." He praised Waksman's "extraordinary fore-sight and rationally conceived planning," which had "systematically explored the antagonism and associations between microorganisms for at least 25 years."

He assumed that the award was "based on this long and arduous, rea-soned investigation quite as much as upon the fortunate chance by which Dr. Waksman finally recognized streptomycin." He did not "presume to judge the intimate relationship" between Waksman and Schatz, although he had not been "favorably impressed" by Schatz's "conduct throughout the whole affair."

Then Mudd suggested a compromise. "The fact that at least in America Dr. Waksman's award was publicized as 'for the discovery of streptomy-cin' in my opinion over-emphasizes the happy event at the expense of the long-term rational effort and opens the way to the kind of criticism Dr. Schatz's associates have made." He added, respectfully, "If, as I assume to be the case, the Nobel Committee takes into account the whole course of the scientist's investigations as well as the happy and perhaps sometimes accidental dramatic benefits that result from this, I think it would be beneficial to make this clear in announcing the award." Mudd sent a copy to Waksman, who replied immediately.

"I certainly would have been very happy if the Nobel committee had taken into account my whole past work rather than the particular instance of the isolation of streptomycin," he wrote. It was unfortunate, but "quite inevitable." Feldman and Hinshaw, he said, rather than Schatz, should have been considered. "The latter was just an assistant, who deserves no greater share in the program of antibiotics than many of my other students."

The Nobel Committee scrambled to alter its original citation. Dr. Arvid Wallgren of the Royal Caroline Institute, who was due to introduce Waksman at the ceremonies in December, was asked by the committee to investigate Schatz's background. Time was short, and he turned to Waksman for help. Waksman was, after all, the de facto Nobel Prize winner.

On November 6, Wallgren wrote Waksman asking for his comments on the "data" he had received from Dr. Reinthaler at the National Agricultural College. "Of course, there is no question of any influence of the contents of the received letter [from Schatz's college] on the opinion of the Caroline Institute," he assured Waksman. But he had a problem. "In my speech I have to reject this attack." Wallgren asked Waksman for comments as soon as possible, as he had to deliver his draft speech in two weeks.

Waksman, far from recusing himself because of an obvious conflict of interest, was ready, even eager, to oblige. He sent Wallgren a three-page "statement regarding the participation of Dr. Albert Schatz in the discovery of streptomycin." Waksman had originally prepared this document for his 1950 lawsuit with Schatz. In it, he asserted that Schatz had "made no independent discovery ... [and had] merely followed detailed instructions by Dr. Waksman." Schatz had known "very little about the problem to which he was to be assigned when he returned from the army in June 1943" and was "no more entitled to any special consideration" than any of the twenty or more other graduate students and assistants who had helped in the solution of the problem of streptomycin.

The document included a description of the alleged break-in to the laboratory by Schatz's Uncle Joe in 1946, when he "carried off certain valuable documents, for which he can be liable for serious damage." It contained a hearsay report, again from 1946, by the hired researcher for the Rutgers PR Department, that Schatz had once said, "Certainly, I had nothing to do with the practical development of streptomycin." Waksman also accused Schatz of being unstable. He said that the Department of Soil Microbiology had "in its possession" evidence concerning Schatz's behavior in two jobs after he left Rutgers that "when placed in the hands of Brooklyn College might make his position there of doubtful tenure." And he added that Schatz was now at a "farm school" of "rather limited academic standing." (Waksman's report was first written when Schatz was still at Brooklyn College.)

The document warned of possible legal action, apparently either by Rutgers or by Waksman himself. "Dr. Schatz has taken an attitude unworthy of an educated person and especially one for whom Dr. Waksman has done so much. Should he attempt to misinterpret the facts in any way at all in the public press he will be liable of [sic] defamation of character, and he and those associates with him will be held in libel suit."

Waksman ended by saying to Wallgren, "I leave it to you to draw your own conclusions as to the motives prompting Dr. Reinthaler's letter."

In a postscript, Waksman said he was "taking the liberty" of sending a copy of his letter and his statement to Professor Nanna Svartz. She was also a member of the faculty of the Caroline Institute, and someone whom Waksman had first met on his trip to Stockholm in 1946.

Wallgren was happy to bring the matter to a conclusion in time for the award ceremony. He thanked Waksman for relating what he accepted as the true story about Schatz's contribution to the discovery. He also told Waksman the award of the Nobel prize would not be reconsidered.

BEFORE HE COULD collect his prize, Waksman had to deal with one more problem from Schatz. Quite coincidentally, in November 1952, Schatz published a grade school book about microbes with a former fellow professor at Brooklyn College named Sarah Riedman. The book was titled *The Story of Microbes* and was a simple explanation of microscopic life and how it affects us—yeast in food, for example—with sections on how to breed and study microbes at home and references to the pioneers of bacteriology: Louis Pasteur, Joseph Lister, and Robert Koch. The style was friendly and the illustrations childlike.

A brief review in *Newsweek* read, "At one point, however, the book lapses briefly into the self-conscious, tight-lipped type of a scientific report. And well it might. In 1943, a particular microbe which normally grows in the soil was found to produce a chemical effective against bacteria that are not harmed by penicillin. This microbial product was named streptomycin. Not until readers get to the dust jacket do they learn whose research led to the discovery of streptomycin; none other than the co-author, Albert Schatz."

The book came off Harper & Brothers' press in November 1952 and was due to be published on December 3, a week before Waksman was to receive his prize in Stockholm. Waksman heard about the book, and saw it as yet another attempt by Schatz to win recognition for himself as a codiscoverer of streptomycin, an attempt that, in Waksman's view, could still spoil his Nobel ceremony. Waksman was again in a litigious mood. He wanted to sue Schatz, to stop "any recurrence of such harassment."

As he was about to board the plane for Stockholm, Waksman called Russell Watson, asking the lawyer to look at the possibilities of stopping the distribution of the book by threatening to sue Schatz for libel. While trying to cool Waksman's desire to sue, Watson took the request seriously and telephoned Frank MacGregor, the president of Harper.

The publisher's publicity sheet had repeated the description on the dust jacket. "At the age of twenty-three Dr. Schatz was responsible for the research which resulted in the discovery of the miracle drug Streptomycin." Watson asserted that the statement was untrue and libelous and offered to send over facts to back up his assertions. McGregor referred the matter to Harper's lawyers. They called Watson in the middle of December, by which time Waksman had received his prize and had left Sweden.

The lawyers doubted the claim of libel, and Watson did not press the point. Advising Dr. Lewis Webster Jones, the new president of Rutgers, of the conclusion of the affair, Watson wrote that in his judgment Schatz was "a dog yapping at the heels of a great world figure and should be ignored."

SCHATZ'S EFFORTS TO enlist top names to his side, including van Niel and Feldman, had failed. The organizer of the appeal, Kurt Stern, wrote to him, "It just goes to show that few individuals, whether scientists or laymen, are willing to go on record in public statements concerning matters which are of a controversial nature. Old Dr. Berliner was right when he said that scientists are 'a curious mixture of a mimosa and a porcupine.'" Stern asked Schatz what they should do next, but warned that the "response of our 'stars' augurs ill." He suggested that a direct appeal to the king of Sweden was still open.

On December 6, Schatz wrote to the king. He and Waksman had agreed "under oath and publicly" that the discovery had been a joint rather than an individual accomplishment, but the prize had been awarded to "only one," and the one who received a large share of the royalties. Schatz had "signed away his patent rights without personal profit on the joint understanding that all royalties would be used exclusively for the benefit of mankind."

"Since the Nobel prize is awarded for a specific discovery," he continued, "the question arises: By what standards of morality and conscience may one of the two acknowledged co-discoverers presume to accept this honor without recognizing the only other co-discoverer?"

Of course, Schatz was not privy to Wallgren's last-minute efforts, or the fact that the Nobel Committee had already taken care of his objection.

BY THE DAY of the award ceremony in Stockholm, Dr. Wallgren had put together introductory remarks to cover Schatz's complaints and still give the prize to Waksman. The committee now understood that it could not give the prize as announced, "for the discovery" itself, so the prize was given for Waksman's "ingenious, systematic and successful studies of the soil microbes that have led to the discovery of streptomycin, the first antibiotic remedy against tuberculosis."

In his introduction, Wallgren said that Waksman had led a "team" in a "long-term, systematic, and assiduous research by a large groups of workers," an "untiring search" for new antibiotics starting in 1939. Repeating Waksman's exaggerated claims, Wallgren said that "no less than 10,000 different soil microbes had been studied for their antibiotic activity" since the start of the program. The year, 1939, was indeed correct. The figure of 10,000 was one of Waksman's "stories."

In another overstatement, Wallgren said that at the time that Waksman had begun his research, "the word antibiotic had not been coined," and Waksman had "introduced" the word as representing an antibacterial substance. Waksman did not "coin" or "introduce" the word "antibiotic," although he was always happy to have the story repeated. Waksman was certainly the first to use "antibiotic" as a noun in a published context, in his

1945 book on microbial antagonism. That is different from coining the word, however.

To put the Schatz affair to rest—at least for the ceremony—Wallgren deliberately mentioned Schatz by name as "one of those" who had worked with Waksman on the team. He also credited Schatz with the isolation of two strains of actinomycetes that produced streptomycin. But Wallgren stressed that the strains were "identical" to the strain of *S. griseus* "discovered by Dr. Waksman in 1915." The difference was that the "rediscovered" microbe "was shown to have antibiotic activity." The suggestion was that Schatz had only "rediscovered" Waksman's original microbe, when in reality he had found an entirely new strain of *griseus* that produced streptomycin.

Waksman must have been pleased with Wallgren's introduction. In his own acceptance speech he avoided the delicate matter of the discovery altogether. The word "discovery" was not even in the title of his lecture, "Streptomycin: Background, Isolation, Properties and Utilization."

And his description of the discovery was "summarized briefly." In a 6,113-word lecture, he never mentioned Schatz's isolation of the two strains 18-16 and D-1, nor did he mention Schatz's experiments to test the antibiotics from those strains against the virulent H37Rv tuberculosis germ. Instead, he devoted most of his lecture to the chemical nature of streptomycin, its antibacterial properties, its toxicity, its effect on infections and diseases, and the resistance of bacteria to streptomycin. The name Albert Schatz appeared only in an appendix, as number twelve in a list of nineteen assistants who had helped Waksman in his research over the years. Half of them had not even been at Rutgers when Schatz had discovered streptomycin. That is how Waksman wanted the world to see Albert Schatz.

IN JANUARY, A month after the award ceremony, Schatz received a reply from the Swedish royal court, not from the king himself, of course, but from his private secretary. It read,

Having made Himself acquainted with the contents of your letter as well as of its appendixes, His Majesty has commanded me to bring the following facts to your attention. The Nobel Foundation is a free and independent institution which by no means is submitted to directions from state authorities. The decisions taken by the different organs of the Foundation regarding the award of the Nobel Prizes—in this case by the Council of the Caroline Medico-Surgical Institute—are, according to express instructions, final and thus not liable to alterations by any superior instances.

The preceding will, I trust, have convinced you that your appeal is not of a nature to call for action on the part of His Majesty.

21 · The Drug Harvest

BY 1953, THE ANTIBIOTIC REVOLUTION WAS the driving force behind a rapidly changing pharmaceutical industry. Besides the "miracle cures" and "wonder drugs" like penicillin and streptomycin, a wide range of new medicines flowed from the expanding drug companies—vitamins, hormones, antihistamines, blood plasma extenders, antimalarials, drugs for hypertension. But antibiotics led the way in the industry's transformation. As American companies screened hundreds of thousands of microbes, a Parke, Davis researcher would famously say that they were finding so many candidate antibiotics that they had to install "an IBM machine" (a computer) to keep track of them. In addition to eight American corporations, companies in other countries began harvesting wonder drugs from the soil—three in France, two in England, two in Italy, one in Sweden, and four in Japan. At the Rutgers Department of Soil Microbiology, Waksman and his students found more than a dozen antibiotics, although only two—neomycin and candicidin—would find widespread practical use.

On his world tours, Waksman was honored and feted by those who had experienced relief from meningitis and TB, and he kept a scrapbook at Rutgers on his "Streptomycin Babies," the children who had survived tuberculosis. In addition to the piles of letters thanking Waksman, and rarely Schatz, for the discovery, however, there were also letters of complaint from those who had experienced toxic side effects.

In reality, the sheen had come off streptomycin. The gray-green culture of *S. griseus*, produced by companies in huge steel vats, was susceptible to a

A group of tuberculous Italian children pose with Dr. Waksman and his wife at the Children's Hospital in Ostia. The children were being treated with streptomycin in 1950. (Special Collections and University Archives, Rutgers University Libraries)

virus, an "actinophage," capable of infecting and destroying the streptomycin-producing mold. This problem was quickly solved, but negative clinical effects were a more enduring issue. Although the drug was nontoxic at low dose levels, the higher doses needed to cure tuberculosis continued to cause the side effects first noticed by William Feldman and Corwin Hinshaw in 1945 and then publicized by British researchers. The Mayo team considered toxicity as one of several "limitations" of streptomycin. Tests showed that streptomycin, and its derivative, dihydrostreptomycin, which had reduced toxicity, could attack nerves responsible for hearing and balance. Among the symptoms were a ringing sound in the ears, vertigo, nausea, rash, fever, and nystagmus—a rapid involuntary movement of the eyeballs. In most cases, however, the symptoms "largely disappeared" within sixty to ninety days after the treatment had been stopped, the Mayo team reported.

Eventually, one of the "most serious obstacles" for streptomycin—as with other antibiotics—was the emergence of drug-resistant strains of the disease bacteria. In some cases, doctors had to increase the initial dose by one thousand times to stop the growth of the TB germ.

The solution came from the synthetic drugs. In the early 1940s, Jörgen Lehmann, a Danish scientist living in Sweden, had been working on an idea to stop TB that was, in fact, considerably more elegant than Waksman's plodding method of screening hundreds of cultures from the soil.

Around the same time that Albert Schatz had isolated *A. griseus*, Lehmann had been inspired by a 1940 paper in the journal *Science* reporting that the TB germ seemed to grow at a rapid rate in the presence of the main ingredient of ordinary aspirin, salicylic acid. Lehmann thought that if he could make "a look-alike chemical"—a derivative of aspirin—that had the opposite effect, inhibiting the growth of the TB germ, he might have an agent against TB. The new chemical was para-aminosalicylic acid, known as PAS. But the Swedish medical establishment was skeptical. PAS was indeed capable of inhibiting the TB germ, but it was not as effective as streptomycin.

While Waksman found ready American sponsors in Merck, the Mayo Clinic, and the Commonwealth Fund, Lehmann, in war-torn Europe, could barely find funds for clinical trials. The German doctor Gerhard Domagk, who had isolated prontosil, found a new class of compounds, known as thiosemicarbazones, that also seemed to arrest the growth of tuberculosis. Despite wartime difficulties, Domagk tested derivatives of these compounds and found one, named isoniazid, that was a potent anti-TB agent. By 1949, however, Lehmann had discovered that a combination of streptomycin and PAS worked much better than one drug on its own. In the end, all three drugs, streptomycin, PAS, and isoniazid, would be needed to defeat the TB germ.

One of the first patients to receive this sort of combination therapy was William Feldman. At the end of 1948, after many years of work with H37Rv, he contracted pulmonary tuberculosis and was treated by Corwin Hinshaw with promin, one of the sulfa drugs; streptomycin; and PAS. He made a complete recovery from the "damnable disease" after a year.

THE DRUG COMPANIES now turned their attention to the new so-called broad spectrum antibiotics, which covered a wider range of diseases. The first was chloramphenicol, found by Paul Burkholder, a Yale microbiologist. Searching for microbes beyond America's farmland, Burkholder called his colleagues around the world and asked them to send him a pot of their local soil, and he subsequently isolated around seven thousand actinomycetes, of which nearly two thousand produced effective drugs. From these he chose four that showed an ability to destroy a wide spectrum of germs—Gram-positive and Gram-negative. Burkholder's champion was *Streptomyces venezuelae*, which he had received from a colleague in Venezuela. It proved its worth during a typhus epidemic on the Peru-Bolivia border in 1947.

With the new antibiotics, the drug companies sought exclusive patents. The thirteen companies producing penicillin, for which there was no patent, had made competition so keen that the price had dropped from twenty dollars for one hundred thousand units in 1943 to four and a half cents in 1950. The downward trend was similar for the eleven companies producing streptomycin. In March 1950, John McKeen, the president of Pfizer, observed, "If you want to lose your shirt in a hurry start making penicillin and streptomycin." Streptomycin was characterized as "distress-merchandise," with production running at 200 to 300 percent above domestic demand.

But the antibiotic hunters, dazzled by the success of streptomycin—the royalties kept rolling into Rutgers's coffers at more than half a million dollars a year—couldn't bear to give up the chase. *Business Week* summarized the position: "The trouble is—from the competitive point of view—that nobody goes out of business. And that is partly because nobody knows what's going to happen tomorrow. A company which is struggling along now in penicillin may come up with a better way of administering it, or a new way of making it. But there is a bigger reason for everyone wanting to hang on. That is the hope that the scores of researchers working in every company's laboratory can come up with an antibiotic it can patent as its own."

The companies sought the much-envied product patent, successfully argued by Waksman and Merck. Product patents were becoming more important than process patents because they gave the company the right to exclude the competition, even in derivatives. In the case of streptomycin, the product patent turned out to cover four chemically different compounds, and even covered dihydrostreptomycin, because to make the new drug you had to start with the old one.

At the beginning of the 1950s, hundreds of antibiotic patent applications were sailing through the Patent Office. (More than six hundred antibiotic patents were issued through 1956.) This flurry would lead to new patent law. The large-scale industrial screening of microbes was adding pressure on the Patent Office to change its original standard for an invention. The courts had previously required inventions to show a "flash of genius"—something inspired and out of the ordinary—in order for a patent to be granted. The essential distinction made was between a traditional inventor and a "skillful mechanic." In the antibiotic revolution, Fleming had certainly had a flash of genius when he had discovered penicillin, and the discovery of streptomycin had been made in the old-fashioned way—by one indefatigable researcher in a basement lab.

Now the big companies engaged in massive screenings of tens of thousands of cultures involving millions in industrial investment. The company researchers could hardly be viewed as genius inventors in the old sense, but they were certainly "skillful mechanics." As William Kingston, of the School of Business at Trinity College Dublin, has written, "a new law which would frankly recognize the change from individuals to investment as the source of what is to be protected, would have needed an amendment to the Constitution. Since this was out of the question, change could only be made in a way that forced the reality of invention by investment into the pretence of invention by individuals."

To accommodate this need, the U.S. 1952 Patent Act provided that "patentability shall not be denied because of the way in which the invention was made." No more flash of genius. That key phrase was replaced with the "inventive step" or "non-obviousness test." A patent was now be granted for something that was new and was "not obvious to one skilled in the relevant art."

In the case of antibiotics, if a company tested a wide variety of microbes from different samples of soil for long enough, it was almost certain to find something patentable. But it would not be "obvious" where to start this treasure hunt. Should it begin, as with Albert Schatz's streptomycin, among the microbes living in a manure pile or on Doris Jones's swab of a chicken's throat? Or in soils on the other side of the world, like Donald Johnstone's coral soils on the Bikini Atoll? At the time, neither of these was an obvious source for medicine. The new "non-obvious" standard was soon adopted internationally and gave the pharmaceutical industry yet another boost.

THE RATE AT which new patents were being granted for substances that were similar in their chemical makeup, and also in their effects, led to another, quite separate trend: companies that sought to keep drug prices high by restricting licenses for their patented products. The result was an antibiotic oligopoly.

Benjamin Duggar, a seventy-two-year-old former professor of botany at the University of Wisconsin and the head of the research department of Lederle Laboratories (American Cyanamid Company), had like Burkholder also been collecting soils from all over the world. He had screened around seven thousand actinomycetes before finding a golden culture, *Streptomyces aureofaciens,* named for its color (after the Latin for "gold"). It produced an antibiotic with a broad spectrum of activity, but it came from a soil sample found relatively close to home—in Columbus, Missouri. He called the drug aureomycin.

Pfizer was also looking for a broad-spectrum antibiotic when it found terramycin. It turned out to be chemically similar to Duggar's aureomycin. The Pfizer chemists fiddled with the structure and produced a more effective product, which the company marketed under a different name, tetracycline. Lederle also discovered tetracycline by the same method and filed its own patent application. Three other companies—Bristol, Squibb, and Upjohn—found tetracycline by another route and filed their patent applications—a total of five.

The Patent Office declared "interference," which usually means a long, hard-fought battle to establish which applicant discovered the new invention first and therefore has "priority" and the right to a patent. Instead of fighting it out in court, however, Pfizer and Lederle got together and settled the matter in a backroom deal in which Lederle conceded priority to Pfizer, which then licensed the drug to Lederle and the three other companies, thus cornering the market for tetracycline for the five companies at a fixed price. Tetracycline would soon become the nation's best-selling antibiotic, with sales topping one hundred million dollars. The maneuver did not go unnoticed by the federal government, however.

THE SECOND GENERATION of anti-infective drugs reached the market in the early 1950s, during the Korean War, when the U.S. military became the primary consumer of such drugs. Congress and the federal government started to take notice of the price of drugs produced in America and, if it was too high, looked elsewhere, importing generics from abroad. The "tetracycline five" attracted a U.S. Senate inquiry, and the government brought a criminal antitrust case against the five companies under the Sherman Act. They were found innocent of collusion, but a damning Senate report concluded that the oligopoly had arisen from three factors: patents; the large sums spent on promotion; and "the well-entrenched practice of physicians of dealing with pharmaceuticals by their brand names rather than their generic categorization." Half a century later, drug industry observers would look back at the Senate report and remark how little had changed.

By the end of the 1950s, some companies would be screening fifty thousand cultures a year. The companies began investing in research and development, and would eventually spend more than 50 percent of their recorded profits on R&D. In 1956, eleven principal drug companies owned 500 antibiotic patents, led by Merck with 111. Fifty percent of all antibiotic patents were for penicillin, 100 patents were for streptomycin and dihydrostreptomycin, and 69 were for broad-spectrum antibiotics.

As the search for antibiotics reached a peak in the mid-1960s, researchers had fun with finding new names. A Greek researcher named one zorbamycin, after the book and movie *Zorba the Greek*, and another melinacidin, after Melina Mercouri, the star of *Never on Sunday*. One isolated by an Italian researcher on February 29 was christened lipiarmycin ("leap-year-mycin" with an Italian accent). By the end of the 1970s, the rate of discovery had fallen off, but more than five thousand potential antibiotics had been found, half of them from the *Streptomyces* genus of the actinomycetes.

Under the new rules of the Food and Drug Administration, the drugs were available only with a doctor's prescription, so the companies soon began to target doctors, who did not pay for the drugs they ordered and often did not know how much the drugs cost the patient. Companies such as Merck and Pfizer began combining discovery, patenting, packaging, and selling, creating the new so-called integrated drug company. When introducing aureomycin, American Cyanamid shipped samples to physicians worth about two million dollars. The sudden increase in the number of advertisement pages in the *Journal of the American Medical Association* told the story. In 1949, there were 32 pages; in 1951, 157 pages; and in 1957, 534. Big Pharma was on its way.

22 · The Master's Memoir

BASKING IN HIS INTERNATIONAL FAME, WAKSMAN continued his foreign tours and, in Japan and France, announced that streptomycin patent royalties would be used to set up Waksman Foundations, to give local students fellowships. In 1954, he also published his memoir.

It would be his fourth publishing opportunity to address the historical record, but Waksman was still angry about Schatz's challenge to his authority. Instead of clearing the fog over the discovery of streptomycin, he still wanted to teach Albert Schatz a lesson.

Waksman sent a draft of the chapter in his memoir describing the discovery to Russell Watson, who spotted many potential libels in Waksman's portrayal of Schatz as a mere pair of hands. "What are you trying to prove by this chapter?" Watson asked. "The intendment [sic] of the chapter is that his [Schatz's] part in the discovery was scientifically negligible." This clearly "opened the door for libel action." Although it was impossible to say whether Schatz would go that far, in Watson's view Schatz was "egotistical, abnormal, possessed of delusions of grandeur and . . . financially able to prosecute a libel action." He could charge that Waksman had unduly minimized his part in the discovery.

"It is impossible to forecast the result of such an action," Watson continued. The original case against Waksman had been heard by an experienced, impartial trial judge, but a libel case would be tried by a judge and jury, and "the outcome would be unpredictable." Schatz could bring up such tricky areas as "the financial arrangement" between Waksman and the university, about which "the reading public would be curious."

Watson also wondered why Waksman kept bringing up the story of the stolen notebook, supposedly taken by Uncle Joe in May 1946 but, according to Watson's brother and law firm partner, Dudley Watson, not stolen at all—they had been in the possession of Merck's lawyers at the time of the alleged theft. Russell Watson insisted that the "missing page" from Schatz's notebook was "insignificant."

The entire chapter about the Schatz case, he commented, would certainly make the book more controversial and increase its sales, but he reminded Waksman that any advice on this matter from his publisher, Simon & Schuster, should be seen as what it was: biased in favor of the marketplace.

In conclusion, Watson said that the various facts of the Schatz case and the proposed chapter were too numerous, and he suggested a meeting. There is no public record of such an event, but Waksman would eventually cut out all references to Schatz by name and removed the parts that Watson had criticized.

MY LIFE WITH THE MICROBES was published by Simon and Schuster in 1954. There was no index, and no references to scientific publications. In discussing the discovery of antibiotics in his laboratory, Waksman mentioned the help of many "assistants," and the discovery of streptomycin itself was described in seventy-five words.

"On August 23, 1943," Waksman wrote, "we isolated a culture of an organism, long known to me, *Streptomyces griseus*. This culture was found, by the methods developed for the production of streptothricin, to produce a similar antibiotic, which we designated streptomycin, a name coined in the laboratory the previous January. Further tests carried out in our laboratory and in the laboratories of Merck & Co, proved it to be a highly desirable substance with potential chemotherapeutic properties."

In the antibiotics program, Waksman said, he had been assisted by nearly fifty graduate students. "They were the fingers of my hand . . . This teamwork might be compared to that of an orchestra, with the conductor leading and assigning the task to each member, none of which [*sic*] would have produced any symphony otherwise." He was not the kind of conductor who picked out his first violin for special recognition and applause. "To name only a few would be a disrespect to others," he wrote.

FOR THE NEXT two decades, Waksman continued to promote himself and streptomycin, at home and abroad. New editions of his autobiography were published in Britain (1958) and Japan (1975, posthumously), still with no mention of Schatz. He wrote and edited several more books on streptomycin and on the actinomycetes. Simon and Schuster turned down his next book, *The Conquest of Tuberculosis*, because, it said, it could not find a market for it.

The book was eventually published in 1964 by the University of California Press, which insisted on an index. Schatz's name appeared six times, in journal references. In telling the story of the discovery, Waksman once again repeated his parable of the sick chicken. The chicken had been responsible for the isolation of streptomycin, not Schatz.

23 · The Copied Notebooks

AMONG THE EVENTS FROM THE SPRING and early summer of 1954 recorded on world-history Web sites are the release of Elvis Presley's first hit single, "That's All Right"; the New York Yankee Joe DiMaggio's marriage to Marilyn Monroe; the sale of the first TV dinners; and the fruition of Selman Waksman's personal dream, the opening of the first Institute of Microbiology, on the Rutgers campus. Waksman's Rutgers monument was a $3.5 million neo-Georgian structure with a gleaming white clock tower, and it was built principally from the streptomycin royalties paid to the Rutgers Foundation.

On opening day, *This Week*, the nationally syndicated magazine, sent the novelist and playwright A. E. Hotchner to be with Waksman as he stepped out of his "modest Chevrolet" and walked up the steps of the institute for the first time. Before entering through the massive front door, Waksman turned, scanning the campus where he had spent most of his life, and his "eyes began to tear, but not from the wind." It had taken Waksman "36 years, living on a meager salary, to find the drug which has produced these fabulous earnings. It is believed that this represents the longest stretch of microbiological research in the annals of medicine."

The headline on the article would read, incorrectly, HE TURNED DOWN MILLIONS, and when asked why, Waksman gave his practiced answer: "I'm too busy to be a millionaire." When Hotchner asked why he hadn't chosen security for his old age, instead of giving his money away, Waksman replied, "The future doesn't worry me ... I will tell you why I did it. The work was done here. Rutgers believed in me and supported me ... But above all—you'll forgive if I wave the flag a little—this country has been good to me, and I feel this money should go back to the country and be used for the benefit of all people."

Hotchner followed Waksman into his new office, past the new laboratories, "glistening with ultra-modern equipment, ready to receive experiments that Dr. Waksman believes can one day solve the mysteries of polio, cancer and even the common cold."

After Waksman had hung up his "worn hat and coat" and sat down at his new desk, a young visiting physician came in desperately seeking the latest batch of streptomycin for his five-year-old daughter, suffering from meningitis. After the doctor had gone, Waksman immediately called his "close friends" at Merck & Co. to discuss finding "a new and possibly more potent strain."

Next, two Rutgers officials entered to ask Waksman's permission to establish a museum room in which all his honors and medals would be displayed. "With reluctance, Waksman agreed but he flatly turned down their request that his portrait be hung in the room. 'Not while I live,' he said."

Then the phone rang. It was the Smithsonian Institution in Washington, D.C., calling about the exhibition they were planning of the discovery of antibiotics. A year earlier, they had approached Waksman seeking laboratory artifacts "unique in the discovery and development of the antibiotic streptomycin." In an exchange of letters in 1953, the Smithsonian had assured Waksman that the exhibition would name him as the sole discoverer of streptomycin, "as well as donor of any specimens which you have to donate to our collection."

Waksman replied immediately by letter, offering an assortment of typical glassware—test tubes, pipettes, petri dishes, and glass flasks used for growing the mold. "The only piece of equipment that is original is our little shaking machine [to shake the microbe cultures during growth] which has since become the model for all shaking machines throughout the world in the screening program for antibiotics," he wrote. It consisted of a dozen glass flasks on a metal bed that was vibrated by an electric motor.

He also happened to have a small amount of the first streptomycin produced in his laboratory, a culture of *S. griseus*, and "numerous photographs of the culture and the antibiotic, as well as the various publications, scientific papers, books, all contributing to the story of the isolation of streptomycin."

As an afterthought, he added a handwritten note at the bottom of the typed letter: "How about notebook pages, etc.?" The Smithsonian had selected three items: the shaking machine, a vial of streptomycin, and *"the original notes books and/or pages kept during the initial investigations which resulted in the discovery of streptomycin* [emphasis added]." The museum also asked Waksman for a photograph.

Eager to supply as much as he could, Waksman had packed up the shaking machine and a platinum needle used in "our early work on actinomycetes which led to the isolation of streptomycin." He sent three vials, one of the "first true" antibiotic, actinomycin; another of streptothricin, "the first water-soluble, basic substance active against Gram-positive and Gram-negative bacteria"; and an example of the standard vials sent out to the makers of streptomycin and to the Food and Drug Administration. He sent four photographs of himself, "two of which are informal and two of which are formal. The larger formal photograph was taken about the time that streptomycin was discovered. You may select from these photographs whichever you would like to use."

He told the Smithsonian that he had three of his own laboratory notebooks, with relevant pages marked. He was careful not to say that they were "the original notes," as the museum had requested; instead he wrote that they "comprise my various notes dealing with the production and isolation of antibiotics leading to the discovery of streptomycin." Because of their "extreme value," Waksman proposed sending them by special registered mail. These might take some time, he warned, because they had not yet decided whether to have photostats made of the pages before sending them out.

The Smithsonian wanted Waksman to choose "the most significant notebook" for the collection. Meanwhile, they had selected one of the photographs and returned the other three. There was no hurry to send the items; the antibiotic collection would not be ready "for some time yet."

Waksman had a problem. The "most significant" laboratory notebook dealing with *"investigations which resulted in the discovery of streptomycin,"* as the museum had requested, was not his, but Schatz's. And the most important experiment in Schatz's notebook was Experiment 11. So Waksman told one of his stories.

He wrote the Smithsonian, "It occurred to me that rather than send you the most important notebook in our collection (since there is always the danger that it might be lost somehow) I have selected the four original experiments which deal with the isolation and production of streptomycin.

"I have re-copied these experiments from my notebook on paper very similar to that used in my notebook and in my own handwriting and as close as possible to the way the data are presented in the notebook. As a matter of fact, the data will be somewhat more readable now since the notes in the notebooks are in pencil and frequently somewhat smudged after constant usage over the past years."

Waksman duly copied four pages of his experiments from his own 1943 notebook, in pencil just like the originals. The first two of these pages— Experiments 55 and 56 (September 15, 1943)—came a full three weeks after Schatz had begun Experiment 11. Waksman's experiments dealt with the type of nutrient used in producing several antibiotics, including those produced by Schatz's D-1 and 18-16 strains, and how six of them had tested against known potentially harmful germs, including *E. coli*.

As he was copying, Waksman made one notation that was not on the original. At the bottom of the page for Experiment 55, he added a "post-script." It read, "D-1 and 18-16 were the two streptomycin-producing cultures, D-1 being the culture isolated from a chicken's throat and 18-16, two days later, from soil."

In addition, Waksman prepared a four-page summary of his life's work on microorganisms. He mentioned his various books, how he had been encouraged by the work of René Dubos in 1939, how he had invented the word "antibiotic," and then the three antibiotics that had been discovered in his laboratory: actinomycin, streptothricin, and streptomycin. In each case, he mentioned, but did not identify, the "numerous assistants" who had helped him.

"The progress of these investigations," Waksman conceded, "was recorded in a number of notebooks. Some of these books are my own and are in my own handwriting. Three of them, marked 'Antagonistic Studies,' have been selected because they contain most of the data on the work which tells the story of the production and isolation of antibiotics leading to the discovery of streptomycin in 1943." Schatz was mentioned by name only in the final paragraph, where he wrote, "Other records, usually more detailed and supplementing my own, are found in several notebooks of my students, notably, H. Boyd Woodruff, Elizabeth S. Horning, Maurice Welsch, Elizabeth Bugie, Albert Schatz, and H. Christine Reilly." All these notebooks were available for anyone to see in the files of the Department of Soil Microbiology, he said, and would later be deposited in the Museum of the Institute of Microbiology—which was being erected in his new building.

The Smithsonian exhibition was eventually installed and opened to the public. Titled "Antibiotics: The Wonder Drugs," it listed five discoverers: Alexander Fleming, penicillin, 1929; Selman Waksman, streptomycin, 1944; Paul Burkholder, Chloromycetin [a trade name of chloramphenicol], 1947; Benjamin Duggar, aureomycin, 1948; and A. C. Finlay et al., terramycin, 1950.

THE RUTGERS PUBLIC relations team saw another opportunity to promote its first and only Nobel prizewinner. On July 1, 1953, Rutgers had announced that "historic hand-written notes" and pieces of lab equipment "relating to the discovery of streptomycin" had been presented to the Smithsonian by Dr. Waksman. The objects had "played significant roles" in the development of streptomycin. The notes described "four original experiments dealing with the isolation and production of streptomycin."

The local New Jersey papers ran the press release almost word for word on July 2. Somewhat more circumspect, the *New York Times* gave Waksman the benefit of the doubt as to whether his notes were original. The *Times* story began, "Four pages of handwritten notes describing the original experiments dealing with the isolation . . ."

The items eventually chosen for the exhibition were labeled as follows: "The pages of Dr. Waksman's notebook show how antibiotic properties of *Streptomyces griseus* were discovered from two different samples: 18-16 from the soil of a heavily-manured field; and D-1 from the throat of a chicken; Inoculating needle used by Dr. Waksman to isolate antibiotic-producing molds and to transfer the first strains of *Streptomyces griseus* which produces streptomycin; Dr. Waksman used this shaking (agitating) machine in his research which resulted in the discovery of streptomycin (to allow larger amount of oxygen in the liquid culture)." Albert Schatz was not mentioned.

PART V · The Restoration

24 · Wilderness Years

In an effort to cheer up his nephew after the blow of the Nobel Prize, Uncle Joe campaigned in his local New Jersey branch of the United States Junior Chamber of Commerce for Schatz to be named as one of America's Ten Outstanding Young Men. The campaign worked. The 1953 citation referred to Schatz as "co-discoverer" of streptomycin, and as the author of more than fifty scientific articles and a popular book on microbes. He had demonstrated "his devotion to science by relinquishing, at the age of 26, all personal gain from streptomycin. His intent was that all royalties should go for scientific research. Only when he discovered later some of the profits were going into private pockets did he take steps to rectify the situation." It marked the beginning of a sad odyssey into scientific obscurity. He would never again find a research position in a microbiology lab that was part of an academic institution.

The ever-vigilant Rutgers PR Department spotted a news story about the award ceremony in the local *Bergen Evening Record,* under the headline "Schatz, Streptomycin Discoverer, Is Honored." The PR team cabled the judges. The attribution was a "gross exaggeration." They hoped the judges would not "knowingly lend their name to the circulation of these untruths."

When Waksman heard about the honor, he was finishing his memoir, *My Life with the Microbes.* The award to Schatz was "a sad commentary on the morals and manners of these times," he had written in a draft of his manuscript, prompting Russell Watson to warn Waksman yet again that his comment was a "wide open door" to a libel action.

If Rutgers and Waksman had known that Uncle Joe had vigorously campaigned for his nephew they would probably have been even more upset, seeing it as yet another subversive maneuver to grab the spotlight. But the citation actually made no claims for Schatz beyond the "co-discoverer" status and had fairly summarized the state of Schatz's career following the lawsuit.

After the Nobel, Schatz stayed on for two years at the small National Agricultural College, covering a "wide variety of problems" to do with "soil fertility, plant diseases, hormones, and a way to diagnose and treat multiple sclerosis." These were vital pursuits, but they lacked the kind of focus one might expect from a researcher who had made a big discovery. By then Schatz was receiving about twelve thousand dollars a year after taxes from the streptomycin royalties, allowing him to move from one new passion to the next—the kind of basic research he literally "loved to do."

He campaigned against the desirability of fluoride in drinking water, studied the effect of bacteria on dental caries, and developed an interest in water diving and the paranormal. Schatz was also fascinated by the feeding habits of mosses and lichens, by how they use a mysterious chemical action known as chelation (from the Greek *chela*, meaning "claw" or "pincer"), a mechanism whereby microbes break down the minerals on rock surfaces. He published a series of papers on chelating in the *Proceedings of the Pennsylvania Academy of Sciences*, and he wrote an eight-page paper on copper mosses, which feed on rocks containing copper, for the *Bryologist*, a journal devoted entirely to liverworts and mosses. Schatz wanted to know how these creatures survived on copper, which is toxic to most microorganisms.

Some of this work had industrial application. With a colleague, he even took out a patent on chelating microbes that might be used in the soil to release minerals useful for plant growth. But that ended unhappily when the National Agricultural College thought that it ought to have been the owner of the patent—even though his colleague did not work for the college and most of the work had been done before Schatz had arrived there and the patent never resulted in any royalties. The disagreement could not be resolved, and Schatz was fired.

He applied for several jobs in research labs and was sometimes accepted by more than one at a time, but as he was deciding which position to take, inexplicably the "offer was gone," as Vivian recalled. Schatz saw the hand of Waksman in these sudden withdrawals, but there is no documentation to support that charge. It seems more likely that it can be attributed to his post-lawsuit reputation as a troublemaker. "So all I did was 'intellectualize about [my ideas],' he told a colleague. "The research went on in my head . . . The positions I did get were 'survival positions.'"

The thrust of his work was often against the establishment; that was his nature. As an early green movement activist, he challenged the fertilizer companies to end their dominant funding of soil research, which had led to decreasing attention being paid to natural, organic fertilizers. An article of his on the subject was published in a 1966 issue of *Compost Science,* an organic-farming periodical. He wanted to pursue this attack further, but could not find sufficient funding.

With Uncle Joe, he set up a small publishing company to publish their research on the chelation theory of dental caries and their opposition to fluoridation as a means of fighting tooth decay. Earlier, Schatz had joined the husband of his sister, Elaine, in running a celery farm in New Jersey, but they could not compete with the low price of the California product.

Whenever an opportunity to set the streptomycin record straight presented itself—a new version of Waksman's story in the newspapers or magazines or in a scientific journal, or a new book with related material—

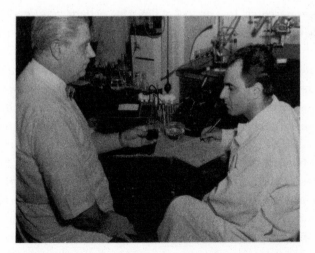

Albert Schatz with Uncle Joe at the National Agricultural College in Doylestown, Pennsylvania. (Courtesy of Vivian Schatz)

Schatz would fire off a letter to the editor, or a colleague, under his Dr. J. J. Martin (Uncle Joe) nom de plume. Two such opportunities arose in 1955.

As "Dr. J. J. Martin," he wrote to William Wightman, a lecturer in the history and philosophy of science at the University of Aberdeen, who had just published *The Growth of Scientific Ideas*. Schatz enclosed details of his discovery of streptomycin—and how the story had been wrongly told.

Thanking "Dr. Martin" for his documentation, Wightman replied, "It is as well for historians to know that neither International Adjudicating Bodies, nor men of science themselves are proof against acts of folly and corruption." The Nobel judges had committed "a double act of folly" in his view, ignoring the codiscoverer and the patent documentation showing what Schatz had done.

But Wightman also did not approve of the Nobel Prize being awarded to *anyone* for the discovery of streptomycin—"a discovery, which, though important in relation to a particular problem of therapy, involved no new scientific principle; being in fact only an extension of the principle first established in the case of penicillin, to which a prize had already been awarded."

Schatz sent another letter under Uncle Joe's name to the Cambridge University professor of animal pathology, W. I. B. Beveridge, who had just published a much acclaimed book on scientific research, *The Art of Scientific Investigation: An Entirely Fresh Approach to the Intellectual Adventure of Scientific Research*. In a review of Beveridge's book, the *New York Times* said that "many of the author's statements deserve to be quoted in every treatise on the psychology and practice of research."

Beveridge replied to "Dr. Martin" that the misallocation of merit in the Schatz case, as "Dr. Martin" had outlined it to him, was "as you say a particularly bad case, but one knows of other such incidents." It showed only, he said, that "some scientists are no better in these matters than other people."

In his book, in a section headed "The Ethics of Research," Beveridge had addressed the question of whose name came first on scientific papers. "Another improper practice which unfortunately is not as rare as one might expect, is for a director of research to annex most of the credit for work which he has only supervised by publishing it under joint authorship with his name first." The aim, he wrote, should be to avoid overlooking the junior person as "merely one of 'and collaborators.'"

Waksman could not be accused of this scientific sin in the Schatz case. He had put Schatz's name first on the two key papers on streptomycin, but when he was challenged later by Schatz, this apparently generous and well-deserved act had in fact no real meaning in Waksman's mind.

Schatz seemed to derive some comfort from continuing to seek justice, as he saw it, through this kind of correspondence. But no one in America, popular journalists or scientists, asked him for his story. Schatz decided to leave the country.

IN 1960, VIVIAN started a Ph.D. at the University of Pennsylvania, and Albert became the bacteriologist at Philadelphia General Hospital. It was a good job, and he had a parking place with his name on it. One day in 1962, he met a visiting professor from the University of Chile, in Santiago, who was looking for an American to help his university organize its science faculties. So Albert, Vivian, and their two daughters moved to Santiago. Vivian became director of the new American School in the Chilean capital of Santiago. It was one of the happiest times of their lives, an opportunity to forget the disappointment of streptomycin.

They stayed in Chile for three years. It was a time of political upheaval, with increasing popular support for Salvador Allende, Chile's future socialist president. "We had Chilean friends who were socialists, and we had to keep that quiet at embassy gatherings," Vivian recalled. But even in their self-imposed exile, it turned out, they could not escape streptomycin and Dr. Waksman.

One day, William Feldman of the Mayo Clinic turned up in Santiago on a tour of South America. After contracting tuberculosis from his experiments, he had suffered through a long period of rehabilitation. Now he had recovered and was enjoying his South American tour. He and Schatz met at the university and chatted about old times. Feldman was surprised to learn that Schatz had been the one who had supplied him his first samples of streptomycin; Dr. Waksman had never mentioned it. And Feldman told Schatz that he had refused to take the money from the royalties that Waksman had allocated to him. It had been a matter of principle not to take rewards from his research, he said.

In the small scientific community that existed in Santiago at the time, the story of Schatz's part in the discovery of streptomycin became widely known. In 1964, he was honored by Chilean doctors and invited to give a speech about his discovery to the medical community, including the Chilean ministers of health and education. It was the first time since his lawsuit that he had been invited to talk about streptomycin to a professional body—and the first time he had described the discovery his 1946 paper to the New York State Association of Public He

Schatz used the occasion to launch a blistering atta Waksman, in all but name. During the past two decades, he said, the story of the discovery of streptomycin had been "enshrouded in an aura of fantasy." Certain "supposed events" had never really occurred, or, at least, "not in the way that has been claimed. Some minor things which did happen, have become exaggerated out of all proportion to their significance. Other really important information has been completely overlooked, distorted, or concealed."

There was a myth, he said, that the discovery of streptomycin had depended, somehow, on the earlier discovery of streptothricin, that it had "pointed the way." In reality, the "research which resulted in the discovery of streptomycin" had been a "logical extrapolation" of the earlier Russian work on antagonistic microbes. "When I began the search for an antibiotic agent effective against tuberculosis, it was the findings of the Soviet investigators, not streptothricin, which gave me confidence that such a substance could be found."

The search for antibiotics at Rutgers had been described as a systematic research when in reality it had been "nothing of the sort," Schatz said. It had involved the "most routine techniques for isolating and testing cultures." There had been no rational basis for choosing one organism over another or for choosing the media in which the cultures had been grown, he said. "No one knows in advance which organism will produce a new and useful antibiotic." The "background information" on actinomycetes in the Rutgers laboratory had not been helpful in knowing which ones to pick. Finally, he said, it had also been a "remarkable coincidence" that the antibiotic he had isolated was effective against Gram-negative organisms, and also the Gram-positive tuberculosis germ.

The attack on Waksman might have gone nowhere if Uncle Joe had not, once again, intervened. He suggested that Schatz send a copy of his lecture to one of his publishing outlets (for his dental caries articles) in Pakistan.

So, two decades after he had isolated *S. griseus*, after many articles by and about Waksman in newspapers, magazines, and scientific journals, after fictional accounts from the Rutgers PR machine, after the lawsuit and the Nobel Prize, Schatz's own account was published not in a journal of microbi· 'gy or medicine in America, or even Europe, but in the *Pakistan ·   eview*, in Lahore. If Waksman saw a copy, he never mentioned i·   .eived no attention in the American or European media.

In 1965, Schatz and his family returned to America at the end of his contract, and Schatz accepted a job at Washington University in St. Louis, teaching science education. But in Missouri, as elsewhere, Schatz could not escape Waksman's shadow. Before he could start teaching, the university required Schatz to have a test for tuberculosis. The test was positive. It didn't mean that the disease would develop, and he had no idea where he had picked up the germ, but it was a stark reminder of his streptomycin work in the basement laboratory. He would always wonder whether it was there that he had picked up the germ.

It was also the year when the patent on streptomycin expired, and another opportunity for the Rutgers PR Department to praise Waksman for his discovery. The *Passaic Herald-News*, faithful to its local hero, reminded its readers that they should also consider the forgotten "co-discoverer." Under the headline "Great Boon, Sad Story," the paper said that it was "unfortunate" that Rutgers "saw fit only to mention" Dr. Waksman.

To celebrate the twenty years of the streptomycin patent and twenty-five years of antibiotics from his lab, Waksman himself wrote an eight-page article titled "A Quarter Century of the Antibiotic Era," for the American Society for Microbiology. In describing *his* discovery, Waksman acknowledged the help of experts from Merck, Pfizer, and Squibb, but Schatz's name did not appear, not even in the referenced scientific papers.

ON WAKSMAN'S EIGHTIETH birthday, in 1968, Rutgers held a celebration in his honor. To mark the occasion, the university produced a book of forty-eight "selected" scientific articles covering his distinguished career. A tribute from his former student Boyd Woodruff concluded, "As a result of the conquering of the scourge of tuberculosis, the accolades of children alive because of his discoveries, the gratitude of parents, the opportunity to dedicate royalties to support new research, all have become the reward of the achievements of a lifetime." The two most important papers of Waksman's career—the announcement of the discovery of streptomycin in 1944, and the report on its action against the TB germ, also 1944, each with Schatz named as senior author—were not among the forty-nine papers selected. In the 386 pages, apart from other scientific listed as references, Schatz's name appeared only on the last page, in a list with seventy-six other students who had worked under Waksman and been awarded advanced degrees. The book's editor, Boyd Woodruff, pointed out that it included Waksman's acceptance speech at the Nobel Prize ceremony in Stockholm. The speech was titled, "Streptomycin: Background, Isolation, Properties and Utilization." Schatz's two key papers are listed in the references. However, Waksman does not refer to Schatz when discussing the drug's isolation, only in a list of twenty of his students at the end.

On August 16, 1973, Selman Waksman died suddenly of a cerebral hemorrhage on Cape Cod. He was eighty-five. He was buried in the local cemetery at Woods Hole. The *New York Times*, in its obituary, used a new adjective to describe his streptomycin work. He was the "principal discoverer." The other twenty-six "who had worked with him on the search" had been rewarded with a share in the royalties "after a court dispute with one of the students." There was no mention of Albert Schatz by name.

At the memorial service at the Rutgers Institute of Microbiology, Byron Waksman spoke of his father's quiet, ironic natural humor. Max Tishler, the Merck chemist who in the 1940s had helped Waksman devise methods of extracting his antibiotics, admired the success of the links he had forged with industry. Ernst Chain, the chemist who had won the Nobel Prize with Alexander Fleming and Howard Florey for penicillin, spoke passionately about a fellow European's assiduous labors in the New World that had brought him to the top of his profession, and secured him a Nobel Prize.

Other tributes flowed—to Waksman's astonishing productivity: more than 350 technical papers, plus writing, editing, or coauthoring about thirty books, all while directing the studies of his seventy-seven graduate students. Waksman did not belong to the publish-or-perish era of researchers— he had no need to prove himself; he was a master in his field. Yet in the decade from 1940 to 1950, his name was on 113 papers, or nearly one a month. He was the lead author on 87 of them. His major book on microbe antagonism was published, in two editions, in the same decade.

The question arose: How much time did he spend in the laboratory bench—and how much of the work was done by his assistants, like Albert

Schatz? In 1940, when he switched his research to full-time antibiotics, he was spending half a day in the laboratory on the third floor of the administration building, according to his graduate student Boyd Woodruff. Then, after his first antibiotics discoveries, he began to spend more time in his office. By the end of the 1940s, according to another graduate student, Hubert Lechevalier, he was rarely seen in the lab, and had become "strictly a manager of research." Lechevalier concluded, "I suspect that he stopped working in the laboratory rather early in his career but that he relapsed from time to time as he found subjects that really interested him."

Still more tributes to Waksman noted his enthusiasm and passion for science, which he was said to have shared liberally with his adoring students. His discovery of antibiotics had been the crowning achievement of his career. Most reviews of his life emphasized his productive links with industry, characterizing him as a pioneer in what today is known as "technology transfer," the often controversial contractual relationships between universities and business.

A few looked at Waksman's career as a scientist and remarked on his preference for applied over pure science, his concentration on the "systematic development of a few ideas" rather than the pursuit of new ones. In this regard, his style of research was compared with that of his former student René Dubos, who had discovered gramicidin in 1939 and had been a big influence on Waksman's change of direction to antibiotics. Bernard Davis, who had collaborated with Dubos on tuberculosis research in the 1940s and had later become a professor of bacterial physiology at Harvard Medical School, made this comparison in 1990: "I would reinforce the picture of Waksman as primarily a natural historian of the soil, cataloguing the microorganisms found there, and focusing on their taxonomy and their ecological effects. He was not a person with the intellectual restlessness that characterized Dubos. But perhaps for that very reason, he was more patient with a kind of search that had to survive several dead ends before yielding a product with the selective toxicity necessary for chemotherapy." Davis suggested that Waksman's "really important discovery was not streptomycin; it was the principle that a patient, systematic search for useful antibiotics will eventually pay off."

Perhaps the best-considered, and the most concise, comparative assessment also came later from Waksman's former student Hubert Lechevalier. He had worked with Waksman on his antibiotic projects in the late 1940s. In 1948, he had produced neomycin from the actinomycete *A. fradii*, described by Waksman and Roland Curtis in 1916 and named after Waksman's mother, Fradia. Lechevalier described Waksman's antibiotic project in a paper given at a conference on the history of antibiotics sponsored by the American Chemical Society. Lechevalier wrote,

> Naturally Waksman considered that he was chiefly responsible for the discovery of streptomycin since it was the fruit of one of his research programs which had already uncovered some interesting antibiotics such as actinomycin and streptothricin ... He had also been mainly responsible for turning it from a laboratory curiosity into an anti-tubercular agent.
>
> Also, naturally, Albert Schatz considered himself co-discoverer of the drug since he had performed most of the basic laboratory manipulations involved in this discovery, and since his name was on the original paper reporting the discovery of this antibiotic, on several other papers published later, and on the U.S. patent which was eventually issued in 1948. In addition, streptomycin was the subject of his thesis which he defended in 1945.

25 · The English Scientist

AND THERE THE STORY MIGHT HAVE ended if it had not been for the curiosity of a young British lecturer in microbiology at the University of Sheffield. In 1987, Milton Wainwright heard that Rutgers was preparing to celebrate the centenary of the birth of Dr. Selman Waksman, and he was intrigued by stories of the dispute between Waksman and Schatz. As a historian of science with a special interest in antibiotics—he had studied Fleming's discovery of penicillin—he went to America in search of the archives.

Wainwright could take his research only so far, however. At Rutgers, the academic staff told him they had no idea where Schatz was, or indeed if he was still alive. His name did not appear in any recent abstracts of scientific papers, and he was not listed as a member of any scientific clubs or associations. He was not even on the Rutgers alumni mailing list.

What Wainwright found in the Rutgers archives, however, convinced him that Schatz alone had discovered streptomycin and that Waksman had unjustly taken the full credit. In 1988, he wrote up his findings in the *Society for General Microbiology Quarterly*, a British periodical founded in 1945 by Alexander Fleming, among others. "Anyone who reads Schatz's thesis cannot doubt that it was he who made streptomycin a reality," he wrote. At the end of the article, he included a footnote to this "major, if largely overlooked scandal," asking readers for any information on "the whereabouts of Dr. Albert Schatz."

By chance, Waksman's former student Hubert Lechevalier read the

Wainwright article. He notified Wainwright that the word on the microbe grapevine was that Schatz was now teaching a course in science education at Temple University, in Philadelphia, an hour's ride from Rutgers.

For many years now, Schatz had avoided speaking about streptomycin. "I stopped long ago telling people what happened," he had written Doris Jones in 1983. If anyone asked, he told them that "the whole thing is buried in the past and I prefer to leave it there. Then some people think that I take that attitude because I feel guilty, ashamed, etc of what I did. So, whatever I do, I can't win."

When Wainwright asked for an interview, Schatz was nervous, but he decided to see him. For one thing, Wainwright had not been involved; for another, he had already shown by his 1988 publication that he was prepared to look into the matter more deeply than others had done.

In February 1989, Wainwright visited Schatz in his modest two-bedroom row house in Mount Airy, a pleasant suburb of Philadelphia. Over four days, the two scientists faced each other across Schatz's dining room table and recorded their conversation on a bulky cassette tape recorder.

For Schatz, it was an occasion filled with emotions that he struggled to control. Apart from Doris Jones, no one—none of his colleagues, no professors, no students, no writers, and certainly not a historian of science— had ever asked him to tell his whole story.

Schatz was entering his seventieth year, but his memory was sharp, and he had assembled for his visitor his personal archive of scientific papers, letters, and newspaper clippings, even photographs of his childhood and his courtship of Vivian. Vivian was also present at the recording session, providing her own recollections.

Slowly and painstakingly, Schatz reconstructed what had happened in the basement laboratory on the Rutgers campus almost half a century earlier. It was just what Wainwright had hoped for, and Schatz obviously found the session therapeutic, a chance to unburden painful memories.

"I can't convey to you what it means to have you come here to ask me these questions," Schatz began. "Do you understand? Nobody has ever done this, nobody in science."

"Let me say, I'm being a bit selfish in coming here," Wainwright replied in his soft, Northern English accent. "I am after the story. I am after the facts."

Albert Schatz interviewed by Milton Wainwright at Schatz's home in Philadelphia, February 1987. (Courtesy Vivian Schatz)

Schatz paused. "But how many have been after the story and written the facts without ever talking to me?" he asked.

"Well, I couldn't do that—" Wainwright began again, but Schatz interrupted him.

"You *are* the first individual in science in forty-five years who has ever expressed a serious interest in finding out what happened from my point of view. You *are* the only one who has expressed an interest in anything that I might have to say."

"I can't understand that," Wainwright said. "It reflects so badly on those scientists."

Wainwright wanted more than just the facts. "It's very difficult," he continued. "I feel I'm being very cold here asking these questions. The way I feel about it sounds a little corny, [but] I cannot believe the injustice. I suppose there's a bit of a rebel in me. I don't like the world to go on without this being known . . ."

It was a strenuous interview. When they came to the Nobel Prize, Schatz could no longer control his emotions.

"So, what did it mean to you—the award of the Nobel Prize to Waksman?" Wainwright asked.

There was a long pause on the tape. Then, softly, Schatz started to sob.

Then his anguish grew and he let out a long howl. At that point, Wainwright relented and turned off the tape.

When Wainwright turned the machine on again, Schatz had recovered his composure.

"The Nobel Prize," he started slowly, "was a denial of my entity as a human being. It was a denial of what my wife and children deserved, of what my mother and father deserved, my uncle, who helped me, who worked in a mattress factory, who worked in an A & P [supermarket] and in a butcher's shop to put himself through dental school."

"The Nobel Prize," he said, "[should have acknowledged] all those people and what they did . . . I'm glad you asked that question. I don't resent it . . . Don't worry about offending me or hurting me . . . I would never have told you this—if you had not asked."

Wainwright searched for something to say that would comfort the older man. "I can appreciate this because I came from the same [working-class] background," he said. "It means something to be able to say to my brother, to my son, hey, I did it. It means something.

"I suppose the only thing I can say . . . is that a lot of people were saved by your contribution. That's the thing you've got to concentrate on . . . In reality, all those kids with [tubercular] meningitis who are now alive because of you. Their families didn't suffer because of you . . . In the final analysis, when you get to wherever one goes, that's what adds up."

Before their meeting was over, Wainwright wanted to know how Schatz felt about Waksman.

"There was a time I really hated him," Schatz said. "I would have killed him if I'd had an opportunity to do so."

He paused, then added, "After a while I guess I just got numb. I tried to bury the whole thing by not talking about it, but I couldn't. Even after he died, he was still there. He was like the weather: He was always there."

AT RUTGERS, THE professor of microbiology, Douglas Eveleigh, who happens to also be British, invited Schatz to visit his old university and give a seminar to students and faculty at the College of Agriculture. Schatz wanted to go, but he did not know if he could stand the level of emotion, on all sides. He suspected, correctly, that there were people at Rutgers who still regarded his lawsuit as an abominable act.

He wrote to Wainwright, "If I accepted I would have to face, encounter, re-live, make peace with, a part of my life that I (and Vivian) have literally blacked out. At first, I was not confident that, emotionally, I would be able to go. But Viv and I talked it over and we decided we would go together."

Schatz called back and accepted, but another three years would go by before the official invitation to return to Rutgers finally came.

IN 1990, WAINWRIGHT published the first popular account of Schatz's work in a short book, titled *Miracle Cure: The Story of Penicillin and the Golden Age of Antibiotics*. It concentrated on new archive material on the discovery of penicillin, but also on what he had been able to find out about streptomycin. There was "no escaping the fact that Waksman reneged on the agreement that he had with Schatz concerning streptomycin royalties," Wainwright concluded. Waksman had also ignored the fact that both the patent documents and the contents of Schatz's thesis proved that Schatz was indeed entitled to be regarded as one of streptomycin's discoverers. "It is all the more tragic then that Waksman could not find the humility to offer Schatz his fair share of both the rewards and the glory that the co-discoverer of streptomycin rightly deserved."

"One of the annoying things," Wainwright later told an interviewer from the Canadian Broadcasting Company, "is that any attempt to redress these problems in favor of Schatz is met with a blanket, a wall. I recently attempted to write an account of the Schatz case, and submitted it to medical journals and so on, and the reviewers have taken a very obstructive line on this. They said that I'm being anecdotal and I'm trying to forward the Schatz case after all this time. My feeling is quite straightforward, that I think Schatz's case stands up. It's very easy to corroborate everything he said, and he's missed out on this opportunity, this great opportunity, to share in the Nobel prize."

26 · A Medal

IN THE WINTER OF 1992, RUTGERS planned another streptomycin celebration, this time for the upcoming fiftieth anniversary of the drug's discovery. As part of the festivities, *Rutgers Magazine*, the alumni journal, ran a three-page article titled "A Nobel Quest." Despite Milton Wainwright's efforts, full honors for the discovery again went to Dr. Waksman. The professor had "found a cure for tuberculosis right under his feet," and "success" had also come to him via the now-infamous sick chicken, which, the article said, had been carrying in its gullet *A. griseus*, a microbe that Dr. Waksman had "identified back in 1915." Waksman "and his researchers" had isolated the strain that produced streptomycin. There was no mention of Albert Schatz, or Doris Jones, or the lawsuit, or Wainwright's two articles, or his book, or a new book *Tuberculosis: The Greatest Story Never Told* by a British physician, Frank Ryan, that included and expanded on Wainwright's original work.

Schatz had come to expect nothing different from Rutgers. Selman Waksman had been dead for almost twenty years, but his ghost lived on, thanks to the Rutgers media team. Schatz immediately wrote three letters to the editor of *Rutgers Magazine*, one after the other over three days in February 1993, complaining about the "egregiously erroneous information" about the discovery and requesting a right of reply. "I am the one who actually discovered streptomycin," he wrote. He was confident "that you and all others at Rutgers University want this 50th Anniversary celebration . . . to be based on fact rather than on fantasy or fiction." He included his usual list of relevant documents in support of his case.

Lori Chambers, the senior editor of *Rutgers Magazine*, thanked Schatz for alerting the publication to the "errors and omissions." They had been "inadvertent rather than intentional." Chambers was "certainly convinced" by Schatz's documentation that "credit for the discovery should be shared by you and Dr. Waksman." She "empathized" with his "frustration." She could "only apologize . . . especially in light of the fact that you are an alumnus and this is the alumni magazine."

However, there was no question of Schatz writing his version of the discovery. There was no need, Chambers assured him. All would be put right in an editor's note in the magazine's next issue. She invited Schatz to send in a "one-page letter" addressing his concerns.

As the fiftieth anniversary neared, another magazine, this time a more prestigious publication, added to the Waksman myths. The January 1992 issue of *Smithsonian*, published by the Smithsonian Institution, described the resurgence of tuberculosis in America, a situation attributed to, among other things, the Reagan-era cuts in community health services. "How TB Survived Its Own Death to Confront Us" also included a summary of the discovery of streptomycin, credited again to Waksman and the sick chicken. Dr. Waksman had "fortuitously discovered" streptomycin after the chicken had picked up a "strange infection from barnyard dirt."

Schatz wrote to the editor, saying that the article was a "complete distortion of history" and adding, "I know . . . because I discovered streptomycin." Doris Jones also wrote to *Smithsonian*, pointing out her part in the discovery and informing the editor that the chicken in question had not been sick and that she had not isolated the strain from the culture she had found in the chicken's throat. Schatz had.

"Over the years," Jones wrote, "the story of streptomycin's discovery has been terribly garbled. I think the Smithsonian magazine would do a great service if it asked Dr. Schatz to tell his own, accurate and interesting account of his finding . . . Dr. Schatz's role has been largely ignored."

Schatz sent copies of his letters to then–Rutgers president Francis Lawrence.

In a letter to *Smithsonian*, Douglas Eveleigh, the professor of microbiology at Rutgers, also complained about the article, naming Schatz and Jones as the isolators of the chicken strain. The various protests made the difference.

Within a few days, Schatz finally received an invitation, via Professor

Eveleigh, to give a lecture to the Biotechnology Club on the Rutgers campus. "At first he wouldn't go, it was too painful," Vivian recalled. The lecture was in the Waksman Institute of Microbiology now named after him. Schatz had never visited the institute. A professor of entomology and plant pathology, Karl Maramorosch, met Albert and Vivian at the train station and drove them to the old college farm. Vivian distracted Albert by pointing out some familiar fungus on the trees. Schatz hesitated at the door of the institute, with its brass plaque to Waksman, but Maramorosch took him firmly by the arm, brought him inside, and insisted on taking a photo of him beside the bronze bust of his old professor that adorned the hallway. Schatz reluctantly agreed. "Let's get this over with," he said.

The flyer for the lecture invited students to "come hear Dr. Albert Schatz, of streptomycin notoriety, reflect on 50 years of research." Schatz told the story of his army service, of watching wounded soldiers die in the hospital for lack of a cure for their infections, and how he had returned to Rutgers determined to find a new antibiotic. He described how he had volunteered to work with the virulent H37Rv strain and how Dr. Waksman had never visited him in the basement lab. He repeated how Waksman had told Doris Jones that he was too immature to accept fame and that's why he had excluded Schatz from the publicity. He told Waksman's parable of the sick chicken, pointed out the errors, and, to prolonged laughter, said, "I don't know what the normal lifespan of a chicken is, but this sick chicken has been alive for half a century. It's amazing."

The lecture was the first in a series of public appearances at professional microbiology meetings, organized by Professor Eveleigh. One of them was the Selman A. Waksman lecture to the Theobald Smith Society, the New Jersey branch of the American Society for Microbiology. "We got a kick out of that one," said Vivian.

A year later, in 1994, to coincide with the fiftieth anniversary of the announcement of streptomycin, Rutgers University presented Schatz with its top honor, the Rutgers Medal. "The worldwide impact of this discovery is now part of medical history," President Lawrence said. "You, thus, have brought distinction and honor to Rutgers, the State University of New Jersey." OVERLOOKED PIONEER FINALLY GETS HIS DUE, said one local headline. "Schatz receives overdue honors," said the Rutgers student daily. "Rutgers will honor a snubbed drug pioneer," said the *Newark Star-Ledger*, whose reporter roamed the award ceremony, looking for quotes. "Schatz was a pioneer," declared Joachim Messing, the forty-eight-year-old director of the Waksman Institute. "I think one should put all the disputes aside . . . It is just a fact that Waksman and Schatz made very important contributions."

Messing and other faculty members believed there had been no malice on Waksman's part. It was a "matter of miscommunication." But Schatz disagreed. In his view, he told the reporter, Waksman had made "an intentional grab for glory." And he, Schatz, had been "blackballed in American science." He had become known as a litigious character, and unemployable, "because I had rocked the boat." Still other Rutgers faculty members, the reporter discovered, continued to hold Waksman in "such high regard that they pooh-pooh Schatz's claim as the ranting of an ingrate."

Schatz came away with a sense of justice. He had been given the medal at the site of the old college farmyard that had produced the soil that had contained his *griseus* strain. But the story of Selman Waksman had been told so forcefully and successfully over the years that one shiny medal was not going to make much difference. A new article in *The Sciences*, for example, gave sole credit to Waksman. Schatz's latest ally Maramorosch wrote a letter complaining. The two authors of the article replied that they had been unaware of the collaboration. They had taken their cue from a brief survey of the historical references, which showed that the "habit of crediting Waksman as the sole discoverer of the drug was indeed widespread."

They quoted Waksman's former student René Dubos, who had once written, regarding scientific discoveries, that "the cruel law of scientific life . . . [is that] . . . credit goes to the man who convinces the world, not to the man to whom the idea first occurs." The authors added, "It is unfortunate, that salesmanship plays so great a role in the recognition of scientific achievement."

Waksman's son, Byron, added his personal view. There was a different academic etiquette in those days, he wrote. Before World War Two, "scientists who directed laboratory programs of any significance regularly appeared as senior authors on all papers emanating from their laboratories. Waksman was one of the first to wish to give his pupils and younger colleagues greater prominence by placing their names before his at the top of his papers. That fact should not mislead anyone about where the ideas, methods and organization of the program of discovery came from." Byron

Waksman said that he "felt sorry for Schatz who was a victim both of the changing fashions in scientific publication and of his own misapprehension of the relative importance of his role in the research."

IN THIS AMERICAN saga, there was one great American institution that had been mistaken about streptomycin. For more than a quarter century, the curators of the Smithsonian's National Museum of American History had believed the sole discoverer of streptomycin to be Selman Waksman. The museum still maintained the antibiotics exhibition including Waksman's artifacts, donated in 1953, among them the pages copied from Waksman's laboratory notebooks.

After Waksman's death in 1973, Schatz had heard about the exhibition and had written to the Smithsonian, asking if his own contribution had been included, or even if his name was mentioned. The Smithsonian had replied with a list of Waksman's items. No, Schatz's name was not mentioned in this exhibition.

On March 9, 1996, Vivian Schatz clipped a short article from the *Philadelphia Inquirer* under the headline PART OF THE ORIGINAL MOLD OF PENICILLIN DRAWS $35,106. One of Fleming's glass slides containing the mold that produced penicillin had been sold at Sotheby's in London. A handwritten inscription on the back of the slide read, "The mold that makes penicillin. Alexander Fleming." The slide had been bought by Pfizer, fetching double what the auctioneers had expected. Pfizer had outbid the Sydney-based Australian Museum of Applied Arts and Sciences.

Vivian wondered whether the Smithsonian might be interested in the two test tubes from Schatz's original streptomycin experiments that he had sealed, one in 1943 and the other in 1944, in his basement laboratory. They were the first samples of the new drug—preserved by Schatz himself and originally given to his mother. One was a culture of *A. griseus*, and the other was an actual sample of streptomycin, part of a batch that Schatz had prepared for the Mayo Clinic for Feldman and Hinshaw's first TB tests on guinea pigs. Surely, the historic value of these test tubes was equivalent to that of a slide of penicillin once used by Fleming, Vivian thought.

The problem was that the test tubes were in England. Schatz had entrusted them to Milton Wainwright with the idea that Wainwright, with

The two test tubes from Albert Schatz's experiments on display
at the Smithsonian's Museum of American History in
Washington, D.C., in 1996. (Courtesy Milton Wainwright)

connections in the world of microbiology that Schatz had long ago lost, might find a place for them in a museum of science, in America, Britain, or elsewhere.

Rutgers had never sought any items from Schatz, even when it had turned his basement laboratory into a museum of the discovery of streptomycin. That collection was also full of Waksman's artifacts.

In 1989, Wainwright had offered the two test tubes to the Smithsonian. There was "absolutely no doubt about the authenticity of the samples," he had written, adding, "As you know Dr. Schatz's role in the discovery has been neglected." But he had never received a reply.

Wainwright now returned the test tubes, and Schatz himself wrote again. The Smithsonian was finally interested, and Schatz carried the tubes in person to Washington. At the National Museum of American History, they were pleased to have these "real treasures" and "thoroughly enjoyed" his visit, which, as it turned out, set in motion the planning of a new exhibition to replace the original one.

Today, the millions of people who visit the museum each year can see an exhibition in the Science in American Life section. There, among Thomas Edison's lamp, an exhibition on atomic power, and early plastic artifacts in a 1950s American kitchen, is a small display case devoted to medical marvels. The case includes a photo of penicillin going on sale in 1945 and Elvis Presley receiving an injection of Jonas Salk's 1955 polio vaccine. There, also, are Albert Schatz's two test tubes.

Next to the tubes is a picture of Schatz in his long white lab coat, tending to his microbes in his basement laboratory. The caption finally, and officially, tells the simple truth: "Dr. Albert Schatz discovered streptomycin in 1943 when he was a 23-year-old graduate student working with Dr. Selman Waksman at Rutgers University."

The pages copied from Dr. Waksman's notebook for the original exhibition are now in storage and out of sight, where they belong.

AFTERWORD

Albert Schatz died on January 17, 2005, at his home in Philadelphia. The cause was pancreatic cancer. He was eighty-four. In obituaries, the *New York Times* and the *Times* of London referred to him as "co-discoverer" of streptomycin but noted that he had to go to court to establish his title.

In the life span of Selman Waksman and Albert Schatz, medicine had gone through a revolution—from a time when there was no cure for bacterial diseases, to the golden age of antibiotics when there was hope that they might also be used against cancer, into another era of uncertainty when the discovery of new antibiotics could not compete with the new resistant strains of bacteria.

These strains still threaten efforts to control tuberculosis in India, China, Russia, and the former states of the Soviet Union. Worldwide, two million continue to die from tuberculosis every year. The pharmaceutical industry, disillusioned by the increasing cost of discovery, lost interest as new antibiotics were desperately needed.

Today, microbiologists are uncovering a fascinating microbial world, far beyond the one known to Waksman and Schatz in which multicolored microbes excreted toxins into zones of antagonism in petri dishes, killing off their neighbors in a battle for survival. Once considered to be single-cell organisms acting on their own and in mute isolation, with little or no capacity for collective behavior, bacteria turn out to have a well-developed social life. They communicate with one another. In response to environmental changes, when colonies are under stress, they send out signals in the form of tiny chemical molecules—some warning of threats to their food supply, others of a need to take action against sudden changes in temperature and acidity. These signals trigger production of antibiotic weapons.

Scientists are working out how to create artificial environments that produce such responses. They are also looking at ways of manipulating microbe genes, and their focus is still on the Streptomyces family that long ago produced streptomycin. Within the genomes of two members of that family, scientists have found clusters of what they call "sleeping genes," which have been present but not active. The trick, so the scientists believe, is to wake up these genes and put them to work, producing new antibiotics. This new generation of researchers is hopeful, devoted, and persistent, as was Albert Schatz in the summer of 1943, that they can find hundreds of new antibiotics.

Acknowledgments

I am indebted to my longtime friend Robert Goodman, dean of Environmental and Biological Sciences at Rutgers University, for his counsel from start to finish on this book. And I owe special thanks to Milton Wainwright, who launched a reexamination of the streptomycin discovery and who loaned me his unique collection of Waksman-Schatz papers.

If there was a eureka moment in my research for this story, it came courtesy of Professor Douglas Eveleigh, Helen Hoffman, and Tom Frusciano and his team of dedicated archivists in the Special Collections and University Archives, Rutgers University Libraries. Over three years, the team fulfilled repeated and undoubtedly nagging requests for access to the sixty manuscript boxes and other material in the Waksman Papers, and also for the papers of the Rutgers Research and Endowment Foundation. However, one request eluded them: Albert Schatz's lab notebooks, which had gone missing. By chance one day, Erika Gorder, a member of the archivists' team, found them in an unmarked cardboard box. The notebook for the year 1943 documents Experiment 11.

At Rutgers, Professer Eveleigh patiently assisted me through several scientific thickets, and provided me with contacts and copies of documents from his own archive. I am grateful to other past and present faculty members at Rutgers who were generous with their time, insights, and hospitality, especially Hubert and Midge Lechevalier, who enriched the story with their own experiences and provided me with documents from archives unavailable elsewhere. Joan Bennett, Karl Maramorosch, and David Pramer added important insights. Among Selman Waksman's former students, Boyd Woodruff generously recalled his experiences in two long interviews. Don Johnstone recounted his extraordinary discovery as a member of the official team of scientific observers at the U.S. atomic bomb tests at Bikini Atoll in 1946.

Vivian Schatz graciously received me at her home in Philadelphia on several occasions, guided me through the Schatz papers at Temple University, and provided new and important material from Albert Schatz's personal archive. Carl Sigmond expertly recorded our conversations. Byron Waksman kindly recalled his childhood with his parents, and provided invaluable firsthand perspectives. My thanks also to Dorothy Hinshaw Patent for her recollections and for help in finding photographs.

Thanks to Tom Whitehead and his staff for their help with the Albert Schatz papers at the Special Collections Research Center of Temple University, Philadelphia; to Renee Ziemer and Robert Nellis at the Mayo Clinic Historical Unit; to the staff of the American Philosophical Society in Philadelphia; to Jay Viszoki at Merck Archives; to the staff at the Library of Congress Manuscript Division; to Amy Schmidt and the staff of the U.S. National Academy of Sciences in Washington, D.C.; to Ellen Alers at the Smithsonian archives, and Diane Wendt at the National Museum of American History, Washington, D.C. In London, thanks to the staff of the United Kingdom Public Record Office.

The following sources gave permission to use and quote from papers, journals, and books: Byron Waksman for permission to quote from his father's works; Special Collections and University Archives, Rutgers University Libraries, for permission to use material and photographs from the Selman A. Waksman Papers and the papers of the Rutgers Research and Endowment Foundation; Vivian Schatz to quote from papers and to use illustrations from Albert Schatz's personal archives; the Special Collections Research Center of the Temple University Libraries for permission to use material from their archives; Mara Ralston for permission to quote from Doris Jones Ralston's letters to Albert Schatz; Dorothy Hinshaw Patent for permission to quote from H. Corwin Hinshaw's letters.

Several others advised on the science, explained military code names, found obscure references, read early drafts, and helped with translations. In this category, my thanks to Natalia Alexandrova, Diana Frank, John Pringle, Julian Perry Robinson, Annika Savill, Viktor Sokolov, and Duncan Taylor.

My agent, Michael Carlisle, wise and enthusiastic as ever, steered the book into the experienced hands of my publisher, George Gibson, at Walker

& Co. in New York and Bloomsbury in London. I was so fortunate to have the expert advice, care, and attention of Jackie Johnson, my U.S. editor, and Patti Ratchford produced a wonderful cover. Thanks also to the Walker team—Christina Gilbert, Laura Keefe, Peter Miller, Laura Phillips—and to copyeditor Lynn Rapoport. In London, thanks to my editor, Michael Fishwick, and to Anna Simpson.

Nothing in my life is possible without Eleanor Randolph, who was, as always, my adviser and first reader, adding her peerless touches to the manuscript. Victoria again provided invaluable computer expertise. Any mistakes are mine.

NOTES

Several archives in the United States and Britain contain documents used to tell this story. The two most important are the Selman A. Waksman Papers (SAW), Special Collections and University Archives (R-MC 003), Alexander Library, Rutgers University, New Brunswick, and the Albert Schatz Papers (AS), Special Collections, Temple University Libraries, Philadelphia. The Rutgers Special Collections and University Archives also hold Schatz's notebooks and the archive of the Rutgers Research and Endowment Foundation (RREF). The U.S. Library of Congress holds another set of Selman Waksman papers (LOC), Collections of the Manuscript Division, Washington, D.C. Vivian Schatz holds her husband's personal archive (AS personal archive). Other archives include the H. Corwin Hinshaw Papers at the American Philosophical Society (APS), Philadelphia; the Modern Military records at the National Archives (NA), Washington, D.C., and College Park, Maryland; the National Academy of Sciences archives (NAS); the Smithsonian Archives (SA); Washington, D.C.; the streptomycin records at Public Record Office (PRO), Kew, London; and the deliberations of the Royal Caroline Institute, Stockholm. Milton Wainwright collected his own archive (MW). A number of former Rutgers staff have papers. They include H. Boyd Woodruff (HBW), Hubert Lechevalier (HL), and Donald B. Johnstone (DBJ). Of the present Rutgers faculty, Professor Douglas Eveleigh maintains his own papers relating to the discovery of streptomycin.

Epigraph

vi **Complete honesty** W. I. B. Beveridge, *The Art of Scientific Investigation: An Entirely Fresh Approach to the Intellectual Adventure of Scientific Research* (New York: Vintage Books, 1957), 196.

Part I: The Discovery

1. Zones of Antagonism

3 **opened his notebook** Albert Schatz's Ph.D. laboratory notebook, vol. 1, June 1943–February 3, 1944, Special Collections, Rutgers University.

2. The Apprentice and His Master

8 **local vigilante committees** Vivian Schatz, author interview, first interview November 8, 2008, and subsequently 2009 to 2011.

8 **heavy smoking** Stanley Rosoff, "Prologue to America," AS personal archive.

10 **"I want to LIVE"** Albert Schatz, *Hilltop Star*, school newspaper, Passaic (New Jersey) High School, April 9, 1936.

11 **"a mere dot"** Selman A. Waksman, *My Life with the Microbes* (New York: Simon & Schuster, 1954), 17.

11 **"prominent merchant"** Waksman, *My Life*, 25.

11 **he was spoiled** Waksman, *My Life*, 26; Byron Waksman, author interview April 20, 2011. Waksman's early life in Ukraine is based on *My Life*, 26–60.

12 **"flying colors"** Waksman, *My Life*, 60.

12 **"perhaps for the last time"** Waksman, *My Life*, 61.

13 **first ten dollars** Waksman, *My Life*, 74.

14 **"unimaginative bore"** and **"great disappointment"** Waksman, *My Life*, 77.

14 **"worked in a sweatshop"** Byron Waksman, draft memoir sent to author, August 13, 2011.

14 **"finally under the tutelage of a master"** Waksman, *My Life*, 78.

15 **"major scientific interest"** Waksman, *My Life*, 81.

15 **his first academic paper** Selman Waksman, "Bacteria, Actinomycetes, and Fungi of the Soil," paper read by Jacob Lipman at a meeting of the Society of American Bacteriologists, Urbana, Illinois, 1915.

16 **"sent my roots into the soil"** Waksman, *My Life*, 87.

3. The Good Earth

17 **dirty petri dishes** Selman A. Waksman, *My Life with the Microbes* (New York: Simon & Schuster, 1954), 102.

18 **Adrenaline** See Joan Bennett, "Adrenaline and Cherry Trees," *Modern Drug Discovery* 4, no. 12: 47–48, 51.

18 **"suggested the possibility"** Waksman, *My Life*, 106.

19 **"demoralizes the assistants"** Selman Waksman to Jacob Lipman, March 8, 1926, LOC.

19 **"zone is found free"** Selman Waksman and Robert Starkey, "Partial Sterilization of the Soil, Microbiological Activities and Soil Fertility," *Soil Science* 16, no. 3 (1923).

19 **"not pursued further"** S. A. Waksman, *The Antibiotic Era: A History of the Antibiotics and of Their Role in the Conquest of Infectious Diseases and in Other Fields of Human Endeavor* (Tokyo: Waksman Foundation of Japan Inc., 1975), 10–11.

19 **"grand scientific tour"** Waksman, *My Life*, 120–21.

20 **"you are the actinomyces man"** Waksman, *My Life*, 153.

20 **"hole of a troglodyte"** Waksman, *My Life*, 146.

20 **"primarily a soil microbiologist"** Waksman, *My Life*, (London: Robert Hale, 1958), 182.

20 **"too busy"** Waksman, *The Antibiotic Era*, 11.

20 *Enzymes* S. A. Waksman and W. C. Davison, *Enzymes: Properties, Distribution, Methods and Applications* (Baltimore: Williams & Wilkins, 1926), 113.

20 **"antagonism and symbiosis"** Selman Waksman, *The Principles of Soil Microbiology* (Baltimore: Williams & Wilkins, 1927 and 1932), 369–71 and 564.

20 **smaller book** S. A. Waksman and R. L. Starkey, *The Soil and the Microbe* (London: John Wiley, 1931), introduction.

21 **"a little Teutonic"** Charles Renner to David Pramer, June 6, 1988, HL.

21 **"Throw it in the basket"** Renner to Pramer.

22 **defatting hides; debutante balls** Charles Renner to Hubert Lechevalier, June 6, 1988, HL.

22 **one of his graduate researchers** Chester Rhines, "The Persistence of Avian Tubercle Bacilli in Soil and in Association with Soil Microorganisms," *Journal of Bacteriology* 29 (1935): 299–311; "The Relationship of Soil Protozoa to Tubercle Bacilli," ibid., 369–81.

22 **"prepared to take advantage"** Selman A. Waksman, *The Conquest of Tuberculosis* (Berkeley and Los Angeles: University of California Press, 1964), 3; see also S. A. Waksman, "Tenth Anniversary of the Discovery of Streptomycin, the First Chemotherapeutic Agent Found to Be Effective Against Tuberculosis in Humans," *American Review of Tuberculosis* 70 (1954): 1–8.

23 **never mentions this as a factor** For an overview of this period in Waksman's career, see a series of papers by Julius Comroe in "Pay Dirt: The Story

of Streptomycin," *American Review of Respiratory Disease* 117, no. 4 (1978): 773–80.

23 **"hung on spikes"** Renner to Lechevalier.

23 *Bacillus mycoides* J. A. Borudulina, "Interrelations of Soil Actinomycetes and B. Mycoides," *Microbiologia* 4, no. 4 (1935): 561–86.

23 **started to turn his mind** George Luedemann, "Free Spirit of Inquiry," *Actinomycetes* 2, supp. 1 (1991): 2.

23 **antibacterial properties** Alexander Fleming, "Selective Bacteriostasis," *Report of the Proceedings of the Second International Congress for Microbiology* (London: Harrison and Sons, 1937), 33.

23 **"seriously interested"** James P. Martin to Milton Wainwright, March 3, 1987, MW.

24 **"substances which are antagonistic"** Selman Waksman and J. W. Foster, "Associative and Antagonistic Effects of Microorganisms: II. Antagonistic Effect of Microorganisms Grown on Artificial Substrates," *Soil Science* 43, no. 1 (1937).

24 **Of eighty cultures** M. I. Nakhimovskaia, "The Antagonism Between Actinomycetes and Soil Bacteria," *Microbiologia* 6 (1937): 131–57.

24 **gramicidin S** Gramicidin S was discovered by G. F. Gause and N. Brazhnikova when, after screening hundreds of bacteria, they finally found the same one as René Dubos, *Bacillus brevis*. Many consider Dubos unlucky not to have won the Nobel Prize in Medicine for his discovery. He wrote and lectured on environmental and social issues and won the 1969 Pulitzer Prize for nonfiction for his warnings regarding the health regarding the earth.

24 **"one cannot escape the possibility"** N. A. Krassilnikov and A. I. Korenyako, "The Bacterial Substance of the Actinomycetes," *Microbiologia* 8 (1939): 673–85.

4. The Sponsor

27 **not even his deputy** Robert Starkey, deposition, *Schatz v. Waksman*, Superior Court of New Jersey, Chancery Division, Docket C-1261-49, July 18, 1950, 337.

27 **exact distribution** Selman Waksman, "Statement Concerning My Relations with Merck & Co.," undated, and "My Connections with Merck & Co.," March 4, 1970, SAW, box 6, 17.

27 **"chemotherapeutic agents"** Selman Waksman, "My Connections with Merck & Co.," March 4, 1970, SAW, Box 6, 7. See also Randolph Major to Selman Waksman, December 15, 1942, SAW, box 1, 3.

27 **"does not appear practicable"** W. H. Helfand et al., "Wartime Initial Devel-

opment of Penicillin in the United States," in *The History of Antibiotics: A Symposium*, ed. John Parascandola (Madison, WI: American Institute of the History of Pharmacy, 1980), 31.

28 **exclusive right** Randolph Major to Selman Waksman, February 17, 1941, SAW, box 1, 3.

29 **"Command me"** Helfand et al., "Wartime Initial Development," 39.

29 **professor of pomology** Ernest Christ, "A History of the New Jersey Peach," www.njaes.rutgers.edu/peach.

29 **50-50 split** H. L. Russell to A. S. Johnson, November 9, 1937, RREF, Box 8, 3.

30 **"off on the wrong foot"** A. S. Johnson to Philip Brett, November 22, 1939, RREF, box 8, 5.

30 **Waksman's graduate students** H. Boyd Woodruff, "Fifty Years Experience with Actinomycete Ecology," *Actinomycetologica* 3, no. 2 (1989): 79–88.

30 **above the chicken house** H. Boyd Woodruff, "A Soil Microbiologist's Odyssey," *Annual Review of Microbiology* 35, no. 1 (1981): 7, 28.

30 **"great provider"** Charles Renner to Hubert Lechevalier, June 6, 1988, HL.

31 **"drop everything"** Woodruff, "A Soil Microbiologist's Odyssey," 33, 7.

31 **"killing machines"** H. Boyd Woodruff, author interviews, December 3, 2010 and September 21, 2011.

32 **"truly excited"** Woodruff, ibid.

32 **"everything changed"** Woodruff, ibid.

32 **no specific name** Woodruff, ibid.

33 **"eyes and ears"** Woodruff, ibid.

33 **"all going to die"** Woodruff, ibid.

34 **400 cultures** Selman Waksman to A. N. Richards, October 1, 1942, LOC, box 3.

34 **on four microbes** Selman Waksman to A. N. Richards, October 1, 1942, LOC, box 3.

34 **"Great White Father"** Doris Jones, "A Personal Glimpse at the Discovery of Streptomycin," undated, 1960, AS personal archive. See also Hubert Lechevalier, "Selman Waksman, Recollections of a 'Latter Day Student and Associate,'" speech on Waksman centenary, Rutgers, May 19, 1988, 2.

34 **"Jewish intellectual"** Hubert Lechevalier, "The Search for Antibiotics at Rutgers University," in *History of Antibiotics*, 113.

34 **work on . . . clavacin** Selman Waksman to Albert Schatz, July 3, 1942, AS personal archive.

35 **Commonwealth Fund** A. N. Richards to Selman Waksman, December 4, 1942, LOC, box 3.

35 **"more valuable"** Selman Waksman to A. N. Richards, October 1, 1942, LOC, box 3.

36 **"eager for the war to end"** Albert Schatz to Selman Waksman, March 31, 1943, SAW, box 14, 4.

36 **Good Conduct Medal** Col. Joseph Benson, U.S. Army Air Corps, Albert Schatz Honorable Discharge, Miami Beach, Florida, June 15, 1943, AS personal archive.

37 **"in his cloth"** Albert Schatz to Ross Tucker, undated letter, AS personal archive.

5. A Distinguished Visitor

38 **"well-padded bones"** Doris Jones Ralston, "A Personal Glimpse at the Discovery of Streptomycin," undated, 1960, AS personal archive, 4.

38 **"I worship him"** Jones Ralston, ibid., 4.

38 **"offer hope"** Sam Epstein, "Streptomycin Background Material," undated, SAW, box 14, 5.

38 **Byron's letter** Waksman produced quotes from the letter in *My Life with the Microbes* (London: Robert Hale, 1958), 212. He wrote that he received the letter in May 1942, but the letter itself was not found in the archives, and Byron Waksman did not know of its whereabouts. Byron Waksman, author interview, April 20, 2011.

39 **"The time has not come yet"** Waksman, *My Life* (London: Robert Hale, 1958), 212.

39 **"early in 1943"** Waksman, *My Life*, 212.

39 **"bacteriostatic substances"** Selman Waksman's Rutgers expenses for the trip to New York, June 1, 1943, SAW, box 1, 13.

39 **dried cells of the human TB** Selman Waksman to Florence B. Seibert, June 18, 1943, SAW, box 14, 3. See also Seibert to Waksman, June 21, 1943.

39 **sent one of each** Selman Waksman to Florence Seibert, June 23, 1943, SAW, box 14, 3.

40 **"true devotee"** W. I. B. Beveridge, *The Art of Scientific Investigation: An Entirely New Approach to the Intellectual Adventure of Scientific Research* (New York: Vintage Books, 1957), 203.

40 **"playing about"** Beveridge, *Art of Scientific Investigation*, 204.

40 **"silly simple"** Hubert Lechevalier, "Antibiotics at Rutgers," in *The History of Antibiotics*, ed. John Parascandola (Madison, WI: American Institute of the History of Pharmacy, 1980), 120.

42 **"excitement that prevailed"** Samuel Epstein and Beryl Williams, *Miracles from Microbes: The Road to Streptomycin* (New Brunswick: Rutgers University Press, 1946), 139.

43 **"I'm paralyzed"** Jones Ralston, "A Personal Glimpse at the Discovery of Streptomycin," undated, AS personal archive, 6.

43 **"designated as streptomycin"** Selman Waksman to Randolph Major, October 28, 1943, SAW, box 6, 7.

43 **call his discovery streptomycin** Albert Schatz, "The True Story of the Discovery of Streptomycin," *Actinomycetes* 4, no. 2 (August 1993): 32.

44 **childhood in Glasgow** Frank Ryan, *Tuberculosis: The Greatest Story Never Told* (Bromsgrove, UK: Swift, 1992), 225.

44 **"foot in the door"** William Feldman, "Streptomycin: Some Historical Aspects of Its Development as a Chemotherapeutic Agent in Tuberculosis," *American Review of Tuberculosis* 69, no. 6 (1954): 861.

44 **"wasting their time"** Feldman, "Streptomycin," 859–68.

45 **Jacob Joffe** Albert Schatz, lecture to the Biotechnology Club, Rutgers University, April 22, 1993, AS personal archive. See also David Pramer, author interview, March 26, 2011.

46 **heard him complain** Albert Schatz to Peter Lawrence, undated 2002, AS personal archive.

47 **first scientific paper** Albert Schatz, Elizabeth Bugie, and Selman Waksman, "Streptomycin, a Substance Exhibiting Antibiotic Activity Against Gram-Positive and Gram-Negative Bacteria," *Proceedings of the Society for Experimental Biology and Medicine* 55 (1944): 66–69.

6. The Race to Publish

48 **non-pathogenic strain** Corwin Hinshaw to Dr. and Mrs. Howard A. Anderson, September 19, 1989, AS personal archive.

48 **a sample of streptomycin** Selman Waksman to William Feldman, March 1, 1944, SAW, box 6, 4.

48 **four to six guinea pigs** William Feldman to Selman Waksman, March 7, 1944, SAW, box 6, 4.

49 **he had pneumonia** Elizabeth Clark to Vivian Schatz, January 1944, AS personal archive.

49 **only member of the staff** Albert Schatz, lecture, *Chilean Society for Diseases of the Chest and Thorax*, Santiago, Chile, November 5, 1964; *Pakistan Dental Review* 15 (1965), 124–34.

50 **problem of logistics** William Feldman to Selman Waksman, April 27, 1944, SAW, box 6, 4.

50 **"one of the finest labs"** William Feldman to Selman Waksman, April 27, 1944, SAW, box 6, 4.

51 **"any possible confusion"** Selman Waksman to William Feldman, May 8, 1944, SAW, box 6, 4.

51 **Feldman agreed** William Feldman to Selman Waksman, May 11, 1944, SAW, box 6, 4.

51 **"mysterious and delicious"** Corwin Hinshaw, biographical notes, December 15, 1990, APS, Corwin Hinshaw papers, series 4, misc.

53 **"sometime Monday morning"** William Feldman to Selman Waksman, July 1, 1944, SAW, box 6, 4.

53 **they were called back** Boyd Woodruff, author interview, December 3, 2010.

53 **only by code names** William Feldman to Selman Waksman, July 19, 1944, SAW, box 6, 4.

54 **"completely inhibited"** Fordyce Heilman to Selman Waksman, August 8, 1944, SAW, box 6, 4.

54 **"definite" confirmation** William Feldman to Selman Waksman, transcript of phone conversation, September 19, 1944, APS, Corwin Hinshaw papers, series 1, correspondence, box 1.

55 **"it would be proper"** William Feldman to Selman Waksman, September 19, 1944, SAW, box 6, 4.

55 **"quite understood"** Randolph Major to William Feldman, September 28, 1944, SAW, box 6, 4.

56 **Feldman agreed to a delay** William Feldman to Randolph Major, October 10, 1944, SAW, box 6, 4.

56 **"We have with us today"** Selman Waksman, *My Life with the Microbes* (New York: Simon & Schuster, 1954), 208–15.

57 **"six-step"** Selman Waksman, Elizabeth Bugie, and Albert Schatz, "Isolation of Antibiotic Substances from Soil Micro-organisms with Special Reference to Streptothricin and Streptomycin," *Proceedings of the Staff Meetings of the Mayo Clinic* 19 (1944): 537–48.

57 **did not produce a clear zone** Albert Schatz, "Report on Waksman's Evaluation of My Role in the Discovery of Streptomycin," 2001, AS personal archive.

58 **"perhaps some intuition"** Douglas Eveleigh and Carl Schaffner, "Reflections on the 50th Anniversary of the Discovery of Streptomycin," *Society for Industrial Microbiology News* 44, no. 4 (July/August 1994): 177–84. Also given as a paper by Douglas Eveleigh for the New Jersey Experiment Station Series D-01111-02.

58 **footnote was printed** Albert Schatz and Selman A. Waksman, "Effect of Streptomycin and Other Antibiotic Substances upon Mycobacterium Tuber-

culosis and Related Organisms," *Society for Experimental Biology and Medicine* 57 (1944): 247.

7. A Conflict of Interest

60 **bizarre operation** John Marquand to S. Bayne-Jones, "Digest of Information Regarding Axis Activities in the Field of Bacteriological Warfare," January 8, 1943, NA, Modern Military Records, Record Group 175.

61 **"bacteria of every description"** Ibid.

61 **"most promising"** E. B. Fred, "Special Conference Concerning the BW Agents and WRS," memorandum, June 17, 1943, NAS Committees on Biological Warfare, series 1, "War Bureau of Consultants" Committee, box 6.

62 **"most of the burns"** Selman Waksman to Randolph Major, October 28, 1943, SAW, box 6, 7.

62 **"hardly fair"** Ibid.

62 **"quite understood"** Randolph Major to Selman Waksman, November 3, 1943, SAW, box 6, 7.

62 **"concrete information"** and **"feasible"** "Biological Warfare: Report to the Secretary of War by Mr. George Merck, Special Consultant," January 3, 1945, NAS, Committees on Biological Warfare, series 1, "WBC" Committee, box 6.

63 **"special service"** George Merck to Vice Admiral Ross McIntire and Surgeon General, August 11, 1944, Secret, declassified November 9, 1989, NA.

63 **"except to the federal services"** Lewis Weed to Major General Norman Kirk, September 27, 1945, NA.

63 **"correspondence and conversations"** Carl Anderson to Selman Waksman, draft letter, June 19, 1944, SAW, box 6, 7.

63 **became a formal letter** Merck & Co. to Selman Waksman, August 17, 1944, SAW, box 6, 7.

63 **"voluntarily abandoned"** Trustees meeting report on compensation paid to Selman A. Waksman, Rutgers Research and Endowment Foundation, New York, December 1950, SAW, box 14, 7.

64 **memorandum of invention** Photostat copy August 14, 1944, RREF, box 3, 24.

64 **"is described in detail"** It is not possible to tell whether Waksman actually carried out these experiments, or whether they were done by one of his graduate students and he then wrote up the results in his notebook. This happened sometimes. The student who might have done the work was Betty Bugie, but her notebooks have not survived.

64 **"whose interests were profits"** Albert Schatz to Jerome Eisenberg, handwritten memo, February 13, 1950, AS, box 2, 25.

65 "without special permission" Secret memo regarding infant, D. W. Richards to Medical Department, Columbia University, September 27, 1944, SAW, box 6, 4.

65 **results were mixed** Edward Miller to N. Paul Hudson, "Conference on Streptomycin at Merck and Company," November 11, 1944, NA, Modern Military Records, Record Group 175.

66 **avoiding overly optimistic statements** Corwin Hinshaw, "Historical Notes on Earliest Use of Streptomycin in Clinical Tuberculosis," *American Review of Tuberculosis* 70 (1954): 9–14.

67 **"Long term crucial"** Selman A. Waksman, *The Conquest of Tuberculosis* (Berkeley and Los Angeles: University of California Press, 1964), 128.

67 **fifty-four cases of TB** William Feldman and Corwin Hinshaw, "Streptomycin: A Summary of Clinical and Experimental Observations," *Journal of Pediatrics* 28 (1946): 269. A preliminary report on thirty-four cases was published in September 1945. H. C. Hinshaw and W. H. Feldman, "Streptomycin in Treatment of Clinical Tuberculosis: A Preliminary Report," *Proceedings of the Staff Meetings of the Mayo Clinic* 20 (1945): 313.

67 **"no conclusive statements"** Samuel Epstein and Beryl Williams, *Miracles from Microbes: The Road to Streptomycin* (New Brunswick, NJ: Rutgers University Press, 1946), ix.

67 **"Schatz and I have discovered"** Selman Waksman, affidavit in the United States Patent Office, in the application of Selman Waksman and Albert Schatz, serial number 577136, February 9, 1945.

PART II: THE RIFT

8. The Lilac Gardens

72 **Lilac Gardens** Vivian Schatz, author interview, November 8, 2008.

73 **four test tubes** Vivian Schatz, author interview, ibid.

73 **"Each morning"** Albert Schatz to Selman Waksman, March 27, 1945, MW.

73 **finished his thesis** Albert Schatz, "Streptomycin: An Antibiotic Agent Produced by *Actinomyces Griseus*," Ph.D. thesis, Rutgers University, 1945, AS personal archive.

74 **"Certain strains of *Streptomyces griseus*"** Selman Waksman, *Microbial Antagonisms and Antibiotic Substances* (New York: Commonwealth Fund, 1945), 117–23.

74 **new genera, Streptomyces** During the 1920s and 1930s, Selman Waksman and others attempted to reclassify members of the group of microbes known as Actinomycetales, but there were so many different types and forms that these efforts failed. In 1943, Waksman and a colleague, Arthur Henrici, laid down new criteria and five genera were recognized. One of these, *Streptomyces*, included the original genus *Actinomycetes*. Thus, *Actinomyces griseus*, which produced streptomycin, became *Streptomyces griseus*.

74 **"*all* infectious diseases"** J. D. Ratcliff, "Keep Your Eye on Streptomycin," *Liberty Magazine*, June 30, 1945, 24–25 and 72.

75 **"Passaic Youth"** Edward Reardon, "Passaic Youth Discovers Drug That May Stamp Out Dread TB," *Passaic Herald-News*, July 2, 1945, 1.

76 **"Magic Germ Killer"** Mona Gardner, "Magic Germ Killer," *Collier's*, August 18, 1945, 23–25.

76 **thirteen-page review** Selman Waksman and Albert Schatz, "A Review: Streptomycin," *Journal of the American Pharmaceutical Association* 6, no. 11 (1945): 308–21.

76 **atomic bomb** "Ten Important Science Developments of Year," *Science Newsletter*, December 22, 1945, 396.

9. The Parable of the Sick Chicken

78 **"things began to happen"** Selman Waksman, deposition, *Schatz v. Waksman*, Superior Court of New Jersey, Chancery Division, Docket C-1261-49, March 25, 1950, 128.

78 **"began to feel uneasy"** Frank Ryan, *Tuberculosis: The Greatest Story Never Told* (Bromsgrove, UK: Swift, 1992), 204.

78 **"all improvements"** A. S. Johnson to Russell Watson, February 1, 1946, RREF.

79 **replaced with "20"** A. S. Johnson, draft agreement between Selman Waksman and Rutgers Research and Endowment Foundation, March 8, 1946, RREF.

79 **"Think it over"** Jerome Eisenberg, notes on interview with Albert Schatz, February 13, 1950, AS, box 4, 38.

79 **"kill job chances"** Jerome Eisenberg, chronology, 1950, MW.

80 **omission of two key papers** J. J. Martin to Robert Strong, April 30, 1946, SAW, box 14, 3.

80 **contacted Waksman** Robert Strong to Selman Waksman, May 23, 1946, SAW, box 14, 1.

81 **"a footnote in the paper"** Selman Waksman to Robert Strong, May 31, 1946, SAW, box 14, 1.

82 **Schatz explained** Albert Schatz to Howard Huber, May 21, 1946, SAW, box 14, 3.

82 **at the company's request** Selman Waksman to J. F. Gerkens, May 5 or 7 (two dates are given on separate pages), 1946, SAW, box 14, 2.

83 **"confidentially"** Doris Jones, deposition, *Schatz v. Waksman*, Superior Court of New Jersey, Chancery Division, Docket C-1261-49, September 26, 1950, 452. Also in D. Ralston, "A Glimpse at the Discovery of Streptomycin," circa 1960, 1, AS.

83 **senior bacteriologist** Harold Lyall to Selman Waksman, April 2, 1946, SAW, box 14, 8.

84 **"has a mature judgment"** Selman Waksman to Harold Lyall, April 4, 1946, SAW, box 14, 8.

84 **"became incensed"** Albert Schatz to Peter Lawrence and Veronique Mistiaen, undated 2002.

84 **"Schatz claim"** Russell Watson to A. S. Johnson, December 30, 1947, RREF.

84 **"no matter how small"** Albert Schatz to Selman Waksman, May 21, 1946, SAW, box 14, 4.

10. Mold in Their Pockets

86 **sudden rise of infectious diseases** Thomas Parran, "The Control of Tuberculosis in the Americas," *Public Health Reports (1896–1970)* 62, 63, Tuberculosis Control issue, no. 16 (June 6, 1947): 827—833.

87 **black market** The market became the subject of popular fiction, notably Graham Greene's *The Third Man*, which also became a movie.

87 **"in their pockets"** P. D'Arcy Hart to A. Landsborough Thomson, July 15, 1946, PRO, Streptomycin, FD 1/6751.

87 **"well-known to this embassy"** Lord Halifax to Ernest Bevin, February 16, 1946, PRO, Streptomycin, FD 1/6751.

87 **increasingly troubled by the desperate calls** A. N. Richards to Sir Edward Melanby, April 13, 1946, PRO, FD 1/6751.

88 **"early optimism"** Our Medical Correspondent, "Guarded Optimism," *Times* (London), October 2, 1946.

88 **British press** "Name Is Kept Secret in Drug SOS," *Daily Express* (London), November 20, 1946.

88 **"really any justification"** Streptomycin PRO, MH 58/636.

89 **black market for streptomycin** *Washington Star*, April 16, 1946.

89 **Porton scientists** D. Herbert, Porton Biological Defence Report, no. 54, "Streptomycin in the Treatment of Experimental 'L' Infections," September 20, 1945, PRO, DEFE 55/156.

90 **"hit-and-miss"** Sir Jack Drummond to Selman Waksman, February 4, 1947, SAW, Streptomycin (British correspondence), box 6, 1.

90 **Watching the test** Donald Johnstone, author interview, April 16, 2011.

92 **"Scientist Tells"** "Scientist Tells of New Drug," *Philadelphia Evening Bulletin*, May 17, 1947.

92 **"embarrass the rest of the world"** Johnstone, author interview, April 16, 2011.

92 **"something new and better"** "Discoverer of New 'Wonder Drug Blasts Fantastic Rumors About It," *Newark Star-Ledger*, May 26, 1947.

11. Dr. Schatz Goes to Albany

93 **"sense of regret"** Robert Clothier to Albert Schatz, June 24, 1946, RREF, box 1, 13.

93 **"most brilliant student"** Elizabeth Clark to Vivian Schatz, September 25, 1947, AS personal archive.

93 **"a triumph for the drug's discoverer"** *Time*, September 16, 1946.

94 **"sole credibility"** *New Jersey Journal of Pharmacy*, June 1946, AS, box 6.

94 **photograph** Albert Schatz to Selman Waksman, November 26, 1946, and Selman Waksman to Albert Schatz, November 29, 1946, SAW, box 14, 4.

94 **"better get used to it"** Albert Schatz to Selman Waksman, February 3, 1947, SAW, box 14, 4.

95 **"determined effort"** Richard Baldwin, *The Fungus Fighters* (Ithaca, NY, and London: Cornell University Press, 1981), 66.

95 **lacking enough egg incubators** Albert Schatz to Selman Waksman, February 20, 1947, SAW, box 14, 4.

95 **"about a year"** Selman Waksman to Albert Schatz, February 4, 1947, SAW, box 14, 4.

96 **"grow lopsided"** Albert Schatz to Doris Jones, July 24, 1946, SAW, box 14, 4.

96 **Robeson could sing** Photo of Robeson at Livingston Junior High School, Albany, New York, United Press International, May 11, 1947.

97 **"12 good eggs"** Albert Schatz to Selman Waksman, February 20, 1947, SAW, box 14, 4.

97 **"Nothing in science"** "The History of Streptomycin," *New York Association of Public Health Laboratories* 26 (1946): 68, MW.

12. The Five-Hundred-Dollar Check

99 **world's largest penicillin maker** "Merck," *Fortune* (June 1946): 106–11.

99 **all except staphylococcus** "Streptomycin," *Life* (February 4, 1946): 57, medicine section.

100 **sixteen dollars a gram** "Medicine: Streptomycin Wonders," *Time* (September 16, 1946).

100 **proved to be a dud** Jane Stafford, "Is Streptomycin the Atom Bomb in the TB War?" *New York World-Telegram* (February 19, 1947).

100 **Waksman's 20 percent** Rutgers Research and Endowment Foundation, minutes, January 16, 1948, RREF.

100 **check for five hundred dollars** Jerome Eisenberg, chronology, 1950, MW.

101 **"and profit considerably"** Albert Schatz to Selman Waksman, January 25, 1948, SAW, box 14, 4.

101 **"anything I want to"** Albert Schatz to Selman Waksman, February 20, 1948, SAW, box 14, 4.

101 **"Peregrinations"** Albert Schatz to Selman Waksman, January 31, 1948, SAW, box 14, 4.

102 **electrifying teacher** Susan Spath, "Van Niel's Course in General Microbiology," *ASM News* 70, no. 8 (2004): 359–63.

102 **"a special recommendation"** Selman Waksman to Albert Schatz, February 3, 1948, SAW, box 14, 4.

103 **"world famed scientist"** Edward Robert Isaacs, Rutgers News Service, April 14, 1948.

103 **"Winner Takes Life"** Bernard Victor Dryer, "Winner Takes Life," a *Cavalcade of America* radio program prepared and produced by Batten, Barton, Durstine & Osborne, Inc., for E. I. du Pont de Nemours & Co., February 1948, LOC.

104 **offered additional support** Selman Waksman to Albert Schatz, June 2, 1948, SAW, box 14, 4.

104 **"to impose further"** Albert Schatz to Selman Waksman, September 7, 1948, SAW, box 14, 4.

104 **"rich as Croesus"** Albert Schatz to Selman Waksman, September 19, 1948, SAW, box 14, 4.

105 **"most important experiment"** Albert Schatz to Selman Waksman, November 22, 1948, SAW, box 14, 4.

105 "pretty tired of hearing about it" Ritch Lovejoy, Round and About, *Monterey Peninsula Herald*, November 3, 1948.

106 "delighted indeed" Selman Waksman to Albert Schatz, October 14, 1948, SAW, box 14, 4.

13. A Patent That Shaped the World

107 "streptomycin and process of preparation" Selman Waksman and Albert Schatz patent application no. 577,136, February 9, 1945, U.S. Patent no. 2,449,866, granted September 21, 1948, U.S. Patent Office, Washington, D.C.

108 "ten patents that shaped the world" Stacy Jones, "Ten Patents That Shaped the World," *New York Times Magazine*, September 17, 1961.

108 poorly stocked medicine chest Peter Temin, *Taking Your Medicine: Drug Regulation in the United States* (Cambridge, MA: Harvard University Press, 1980), 59.

109 "products of nature" Richard Seth Epstein, "The Isolation and Purification Exception to the General Unpatentability of Products of Nature," *Columbia Science and Technology Review*, January 15, 2003.

109 adrenal glands Joan Bennett, "Adrenaline and Cherry Trees," *Modern Drug Discovery* 4, no. 12 (2001): 47–48, 51.

110 "good ground for a patent" Jon Harkness, "Dicta on Adrenalin(e): Myriad Problems with Learned Hand's Product-of-Nature Pronouncements in Parke, Davis v. Mulford's," 44, http://ssrn.com/abstract=1881193.

110 "cannot be claimed per se" Selman Waksman and Boyd Woodruff, U.S. Patent Application, October 2, 1941, no. 413,324, granted June 19, 1945, no. 2378876. See also *Marcus v. Waksman et al.*, SAW, box 16, 2.

110 "lingering doubt" Selman Waksman, amendment, May 17, 1945, to patent application no. 413,424, October 2, 1941.

111 "failed to detect it" Selman Waksman, response to patent examiner's objections, June 5, 1946, in patent application no. 577,136, February 9, 1945.

111 "many pathogens can grow" Selman Waksman, *Microbial Antagonisms and Antibiotic Substances* (New York: Commonwealth Fund, 1945), 15.

111 "also in soil" M. I. Nakhimovskaia, "Antagonism Among Bacteria," *Microbiologia* 6 (1937): 131–37. Also in Waksman, *Microbial Antagonisms*, 116.

112 "At other times . . . he would say" David Pramer, author interview, March 22, 2011. See also David Pramer, "The Persistence and Biological Effects of Antibiotics in the Soil," New Jersey Agricultural Experiment Station, Rutgers (1957): 221–24.

112 **"particularly pleased"** Russell Watson to B. R. Armour, September 4, 1948, RREF.

112 **"prevent the importation"** Russell Watson to B. R. Armour, ibid.

112 **"Thus, for the first time"** Robert Peck, patent application no. 612,557, August 24, 1945, U. S. Patent Office, Washington, D.C.

113 **as a loan** Albert Schatz to Selman Waksman, September 7, 1948, SAW, box 14, 4.

113 **another $500 check** Selman Waksman to Albert Schatz, October 14, 1948, SAW, box 14, 4.

113 **treat the checks** Selman Waksman to Albert Schatz, November 16, 1948, SAW, box 14, 4.

113 **"what to do with the money"** Albert Schatz to Selman Waksman, November 29, 1948, SAW, box 14, 4.

113 **"a certain sum of money"** Selman Waksman to Albert Schatz, December 2, 1948, SAW, box 14, 4.

114 **his own income tax** Ibid.

PART III: THE CHALLENGE

14. The Letter

118 **"several matters"** Albert Schatz to Selman Waksman, January 22, 1949, SAW, box 14, 4.

120 **"To say that I was amazed"** Selman Waksman to Albert Schatz, January 28, 1949, SAW, box 14, 4.

121 **none of Schatz's business** Hubert Lechevalier to Albert Schatz, February 12, 1993, HL.

124 **a mere pair of hands** Selman Waksman to Albert Schatz, February 8, 1949, SAW, box 14, 4.

124 **Schatz's name** Selman Waksman to J. F. Gerkins, May 7, 1946, SAW, box 14, 2.

15. Choose a Lawyer

126 **"without the name of Schatz"** Selman Waksman to Rutgers Research and Endowment Foundation, memo, February 2, 1949, SAW.

128 **"highly confidential"** Selman Waksman to Chester Stock, May 20, 1949, SAW, box 14, 5.

128 **"request him to leave"** Chester Stock to Selman Waksman, May 21, 1949, SAW, box 14, 5.

129 **"very fond of"** Gilbert Dalldorf to Albert Schatz, May 17, 1949, LOC.

129 **"no skeletons"** Doris Jones to Albert Schatz, February 17, 1949, AS.

130 **"lose your temper"** Seymour Hutner to Albert Schatz, February 28, 1949, AS.

131 **"former Passaic man"** *Passaic Herald-News*, May 5, 1949.

131 **"Choose a lawyer"** Julius Schatz to Albert Schatz, Jerome Eisenberg chronology, June 6, 1949, MW.

131 **"Being laymen"** M. D. Bromberg Associates letter, June 22, 1949, AS personal archive.

132 **"the truth would out"** P. P. Pirone to M. H. Bromberg, June 30, 1949, AS personal archive.

132 **"absolute fact"** Doris Jones to M. D. Bromberg Associates, June 30, 1949, AS personal archive.

133 **minimize the part** Kent Wight to M. D. Bromberg, July 6, 1949, AS.

133 **"happy to give credit"** Boyd Woodruff to M. D. Bromberg, August 12, 1949, AS.

133 **"Malicious"** Selman Waksman, handwritten note on letter from M. D. Bromberg to Elizabeth Clark, June 22, 1949, SAW.

133 **"W and W Sleuthing Agency"** Russell Watson and Selman Waksman, memo and reports, September 1949, SAW, box 14, 6.

135 **"nuisance value"** Jerome Eisenberg, chronology, 1950.

135 **"die is cast"** Seymour Hutner to Milton Wainwright, November 17, 1987, MW.

135 **story about antibiotics** "The Healing Soil," *Time* cover story, November 7, 1949, 70–76.

16. The Road to Court

137 **"extra-marital affairs"** Albert Schatz to Peter Lawrence and Veronique Mistiaen, undated 2002.

137 **"taken for a communist"** Jerome Eisenberg, "Recollections, Schatz v. Waksman et al," 19–39, AS, boxes 3, 34 and 4, 38.

137 **"pendulum swung"** Hubert Lechevalier to Albert Schatz, March 4, 2001, AS personal archive.

137 **"most credible and intelligent"** Eisenberg, "recollections," 21.

138 **"moral torment"** Ibid., 20.

140 **"don't fail to call on me"** Doris Jones to Albert Schatz, March 13, 1950, AS.

140 **"certainly stirred up"** Doris Jones to Albert Schatz, undated, 1950, AS.

141 **"scandalous"** A. J. Goldforb to Selman Waksman, March 14, 1950, SAW, box

14, 2. See also Selman Waksman to Russell Watson, March 16, 1950, SAW, box 14, 2.

141 **"Since this money"** William Steenken to Selman Waksman, March 15, 1950, SAW, box 14, 8.

142 **"never expected nor did I want"** Selman Waksman to Russell Watson, March 21, 1950, SAW, box 14, 2.

17. Under Oath

143 **most interesting depositions** Jerome Eisenberg, notes, *Schatz v. Waksman*, AS, box 3, 4, 1.

143 **"with my own fingers"** Selman Waksman, deposition *Schatz v. Waksman*, Superior Court of New Jersey, Chancery Division, Docket C-1261-49, March 25, 1950, 14.

145 **"It did not produce streptomycin"** Selman Waksman deposition, ibid., 33.

145 **"we observed certain cultures"** Selman Waksman deposition, ibid., 40.

146 **"depends entirely"** Selman Waksman, *Schatz v. Waksman*, 62.

146 **"culture in the flask"** Selman Waksman deposition, ibid., 62.

147 **"could not answer that question"** Selman Waksman deposition, ibid., 66.

148 **one of my bright students** Selman Waksman deposition, ibid., 73.

148 **"NOT TRUE"** Selman Waksman deposition, ibid., 116.

149 **recognized his contribution** Selman Waksman deposition, ibid., 119.

150 **total . . . $350,000** Selman Waksman deposition, ibid., 99.

151 **"Nonsense, never done"** Selman Waksman deposition, ibid., 120.

151 **"they were independent isolations"** Selman Waksman and Albert Schatz, "A review: Streptomycin," *Journal of the American Pharmaceutical Association* 6, no. 11 (1945): 309.

151 **titled *Streptomycin*** Selman Waksman, ed., *Streptomycin* (Baltimore: Williams & Wilkins, 1949), 11.

152 **"Funny business"** Jerome Eisenberg, chronology, 1950, MW.

153 **"Now tell me"** Selman Waksman deposition, ibid., 128.

153 **don't recall who that was** Selman Waksman deposition, ibid., 129.

156 **"insignificant"** Russell Watson to Selman Waksman, February, 10, 1954, SAW, box 14, 6.

156 **could not have been true** Russell Watson to Selman Waksman, ibid.

156 **"Waksman is through"** Albert Schatz to Doris Jones, April 19, 1950, AS.

157 **"a money conscious fool"** Doris Jones to Albert Schatz, undated, 1950, AS.

157–58 **"hoodwink the public"** "Rutgers Is Too Smart for Its Own Good," *Passaic Herald-News*, May 1, 1950.

158 **"no attempt to justify"** Bob Starkey to Selman Waksman, May 10, 1950, SAW, box 14, 2.

159 **"Do you remember what plates"** Jerome Eisenberg, Fred Beaudette deposition, *Schatz v. Waksman*, Superior Court of New Jersey, Chancery Division, Docket C-1261-49, September 26, 1950, 223.

160 **"maybe fifty"** Doris Jones deposition, *Schatz v. Waksman*, Superior Court of New Jersey, Chancery Division, Docket C-1261-49, September 26, 1950, 446.

161 **"get at the true facts"** Doris Jones to Albert Schatz, undated, 1950, AS.

18. The Settlement

164 **close scrutiny** Russell Watson and Dudley Watson, memo, meeting of the Rutgers Board of Trustees, December 15, 1950, 1–5, SAW, box 14, 7.

165 **redistribution of the royalties** Memorandum of Proposed Terms of Settlement of *Schatz v. Waksman* and Rutgers Research and Endowment Foundation Discussed by Russell E. Watson and Messrs. Eisenberg and Libert, December 19, 1950, SAW, box 14, 7.

166 **"role of Maecenas"** Hubert Lechevalier, "The Search for Antibiotics at Rutgers University," in *The History of Antibiotics: A Symposium*, ed. John Parascandola (Madison, WI: American Institute of the History of Pharmacy, 1980), 119.

166 **"an excellent one"** Judge Thomas Schettino, statement by the court, *Schatz v. Waksman*, Superior Court of New Jersey, Docket C-1261-49, December 29, 1950.

167 **"a matter of public record"** Robert Clothier, "Statement" on the settlement, December 29, 1950, in January–February 1951 Faculty Newsletter, Rutgers University.

167 **"long-winded explanations"** "He Finally Gets Credit," Editor's Opinion, *Newark Star-Ledger*, December 30, 1950.

167 **"influenced by the fact"** "Dr. Schatz Wins 3% of Royalty; Named Co-Finder of Streptomycin," *New York Times*, December 30, 1950.

169 **"well hushed-up"** Doris Jones to Albert Schatz, January 11, 1951, MW.

169 **"faithful assistance"** Selman Waksman to Doris Jones, January 5, 1951, MW.

170 **"not unmindful of the fact"** Robert Starkey to Selman Waksman, January 8, 1951, SAW, box 14, 7.

170 **"no direct claim"** Boyd Woodruff to Selman Waksman, January 1, 1951, SAW, box 14, 7.

170 **"best years of my life"** Dale Harris to Selman Waksman, January 10, 1951, SAW, box 14, 7.

170 **"utterly surprised, even amazed"** Corwin Hinshaw to Selman Waksman, January 2, 1951, SAW, box 14, 7.

170 **declined to accept** William Feldman to Selman Waksman, February 13, 1951, SAW, box 14, 7.

171 **"to refuse the small royalty"** Selman Waksman to William Feldman, February 19, 1951, SAW, box 14, 7.

171 **"I am sorry"** Selman Waksman to Walton Geiger, February 19, 1951, SAW, box 14, 7.

171 **"in the hope that"** Russell Watson to Selman Waksman, February 8, 1951, SAW, box 14, 7.

PART IV: THE PRIZE

19. The Road to Stockholm

175 **almost being fired** Raoul Tunley, "He Turned His Back on a Million Dollars," *The American Magazine*, March 1952, 21.

175 **"biggest hoax"** Selman Waksman, "Statement Made by Dr. S. A. Waksman Pertaining to His Connections as Consultant with Industrial Organizations and Various Scientific Institutions," undated, LOC, box 1.

176 **seventh year in a row** Nobelprize.org, Nomination Database for the Nobel Prize in Physiology or Medicine, 1901–1953, accessed at http://www.nobel prize.org/nobel_prizes/nomination/.

178 **deserved a prize** J. P. Strombeck," "Betankande Angaende, Selman Abraham Waksman," Nobel Archives, 1952, Ard. 3:16, AS personal archive.

178 **a more difficult task** Einar Hammersten, "Betankande Angaende, Elizabeth Bugie, Karl Folkers, Albert Schatz, Selman Abraham Waksman, och Oscar Wintersteiner," Nobel Archives, 1952, Ard. III: 4, AS personal archive.

179 **"association of steps" taken** H. Boyd Woodruff, deposition, *Schatz v. Waksman*, Superior Court of New Jersey, Chancery Division, Docket C-1261-49, July 12, 1950, 271.

179 **"only scientific publications"** Professor Hilding Bergstrand, president of the Nobel Committee for Physiology or Medicine, told the Swedish media that the award to Dr. Waksman had been made solely on the basis of scientific publications and that his committee had been satisfied with the papers it had received. "I must, however, point out," said Bergstrand, "that when we pick Nobel Prize winners we do not take into account legal proceedings. We consider exclusively scientific publications concerning the

work of and by prospective candidates." See "Student fann streptomyci-
net tillsammans med Waksman" [Student Found Streptomycin Together
with Waksman], *Goteborgs Handels-Och Sjofarts-Tidning*, October 25,
1952.

178 **to show his data** Corwin Hinshaw to Dr. and Mrs. Howard A. Anderson,
September 19, 1989, AS personal archive.

181 **as the discoverer** Einar Hammersten, "Betankande Angaende, Elizabeth
Bugie, Karl Folkers, Albert Schatz, Selman Abraham Waksman, och Oscar
Wintersteiner," Nobel Archives, August 21, 1952, 10, AS personal archive.

20. "A Dog Yapping at the Heels of a Great World Figure"

182 **citation was specific** George Axelsson, "Waksman Wins Nobel Prize for
Streptomycin Discovery," *New York Times*, October 24, 1952.

183 **"immigrant boy"** Selman Waksman, *My Life with the Microbes* (New York:
Simon & Schuster, 1954), 305.

183 **"bedlam"** Waksman, *My Life*, 305.

183 *New York Times* Axelsson, "Waksman Wins Nobel Prize."

183 **"share in the honor"** "Nobel Prize for Waksman," *Philadelphia Inquirer*,
October 26, 1952.

183 **"entity as a human being"** Albert Schatz, interview by Milton Wainwright,
1989, MW, February 18, 1989.

184 **"amazement"** Elmer Reinthaler to Göran Liljestrand, October 29, 1952, AS,
box 5, 6.

184 **"rightful share"** Albert Schatz "On the award of the Nobel Prize in Physiology or Medicine for 1952," November 1952, MW.

185 **"ungrateful, spoiled, immature child"** Albert Sabin to Elmer Reinthaler,
November 10, 1952, AS and MW.

185 **"junior person"** Maurice Stacey to Elmer Reinthaler, November 24, 1952,
AS and MW.

186 **"ample justification"** C. B. van Niel to Kurt Stern, November 17, 1952, AS
and MW.

186 **"disappointed"** William Feldman to Albert Schatz, November 12, 1952, AS
personal archive.

186 **"distinction is extremely important"** William Feldman to Kurt Stern, November 19, 1952, AS.

187 **"generally regretted"** Hilding Bergstrand and Göran Liljestrand to Elmer
Reinthaler, November 14, 1952, AS, box 5.

187 **"low attempt by little men"** J. C. Hoogerhide to Selman Waksman, November 15, 1942, SAW, box 14, 9.

188 **advised strongly against any litigation** Russell Watson to Selman Waksman, November 21, 1952, SAW, box 14, 9.

188 **"long-term rational effort"** Stuart Mudd to Göran Liljestrand, November 11, 1952, AS, box 5.

189 **"deserves no greater share"** Selman Waksman to Stuart Mudd, November 14, 1952, SAW, box 14, 9.

189 **"I have to reject this attack"** Arvid Wallgren to Selman Waksman, November 6, 1952, LOC.

190 **"draw your own conclusions"** Selman Waksman to Arvid Wallgren, November 11, 1952, AS, box 5.

190 **Wallgren was happy** Arvid Wallgren to Selman Waksman, November 19, 1952, LOC.

190 **"self-conscious, tight-lipped"** *Newsweek*, science section, December 15, 1952.

191 **"dog yapping at the heels"** Russell Watson to Lewis Webster Jones, January 29, 1953, SAW, box 14, 9.

191 **"mimosa and a porcupine"** Kurt Stern to Albert Schatz, November 21, 1952, AS, box 5, 2. Dr. Arnold Berliner was the famous editor of the German *Die Naturwissenschaften*, and his struggle to find well-written, clear, and succinct articles by scientists led him to say that a scientific author should be a cross between a mimosa and a porcupine.

192 **"By what standards of morality"** Albert Schatz to Gustav VI, December 6, 1952, AS.

192 **"ingenious, systematic and successful studies"** Arvid Wallgren, Presentation Speech, Stockholm, December 10, 1952, accessed at http://www.nobel prize.org/nobel_prizes/medicine/laureates/1952/pres.html.

192 **"10,000 different soil microbes"** Hubert Lechevalier to Byron Waksman, May 26, 1981, HL. The figures were "Waksmanesque." Douglas Eveleigh, author interviews. 2008, 9, 11.

193 **That is different** Hubert Lechevalier to Boyd Woodruff, October 19, 1981. See also R. Bentley and J. W. Bennett, "What Is an Antibiotic? Revisited," *Advances in Applied Microbiology* 52 (2003): 303–31. Toward the end of his life Waksman modified his claim, saying he had been the first to "redefine" the word "antibiotic." Selman Waksman, letter to the editor, undated, but sent when he was director of the Rutgers Institute of Microbiology (1954–58), *Actinomycetologica* 24, no. 2, 2010.

193 **"summarized briefly"** Selman Waksman, "Streptomycin: Background, Iso-

lation, Properties and Utilization," lecture, Stockholm, December 12, 1952, www.nobelprize.org/nobel_prizes/. . ./1952/waksman=lecture.

193 **"commanded me to bring"** C. F. Palmstierna to Albert Schatz, January 5, 1953, AS.

21. The Drug Harvest

195 **"IBM machine"** Dana Thomas, "Broader Spectrum, the Wonder Drugs Are Finding New Uses in Commerce and Industry," *Barron's*, November 5, 1956, 3.

196 **"limitations"** William Feldman and H. Corwin Hinshaw, "Streptomycin: A Valuable Anti-Tuberculosis Agent," *British Medical Journal*, January 17, 1948, 91.

196 **ringing sound** Milton Wainwright, *Miracle Cure* (Oxford, UK: Basil Blackwell, 1990), 138.

196 **"most serious obstacles"** Feldman and Hinshaw, "Streptomycin." Also see Davies, "Where Have All," 287–90.

196 **Jörgen Lehmann** Frank Ryan, *Tuberculosis: The Greatest Story Never Told* (Bromsgrove, UK: Swift, 1992), 144–47 and 242–47.

197 **isoniazid** Ryan, *Tuberculosis*, 349–50 and 353–63.

198 **"damnable disease"** William Feldman to Selman Waksman, October 28, 1949, LOC, box 1.

198 **sought exclusive patents** Federal Trade Commission, *Economic Report on Antibiotics Manufacture* (Washington, D.C.: Government Printing Office, 1958), 228.

198 **"If you want to lose your shirt"** John McKeen in *Business Week*, March 26, 1950, 26.

198 **price had dropped** *Business Week*, March 25, 1950, 26.

199 **"pretence of invention"** William Kingston, "Antibiotics, Invention and Innovation," *Research Policy* 29 (2000): 697.

200 **the five companies** See Christopher Harrison, *The Politics of the International Pricing of Prescription Drugs* (Westport, CT: Praeger, 2004).

200 **one hundred million dollars** United States Court of Appeals for the Sixth Circuit, June 16, 1966, 363 F. 2d 757, 4 line 20.

201 **Half a century later** Christopher Scott Harrison, *The Politics of the International Pricing of Prescription Drugs*, 47.

201 **more than 50 percent** "Overview of the U.S. Pharmaceutical Industry: The Competitive Status of the U.S. Pharmaceutical Industry," *National Academy of Sciences* (1983): 7, accessed at http://www.nap.edu/openbook/030903969/hmtl/7.

201 **more than five thousand** S. T. Williams and J. C. Vickers, "The Ecology of Antibiotic Production," *Microbial Ecology* 12 (1986): 43–52.

201 **so-called integrated drug company** Peter Temin, "Technology, Regulation and Market Structure in the Modern Pharmaceutical Industry," *Bell Journal of Economics* 10, no. 2 (1979): 43.

201 **advertisement pages** Federal Trade Commission, *Economic Report*, 13.

22. The Master's Memoir

202 **fourth publishing opportunity** Three books written, edited, or supervised, by Waksman had been published to date: Samuel Epstein and Beryl Williams, *Miracles from Microbes: The Road to Streptomycin* (New Brunswick, NJ: Rutgers University Press, 1946); Selman Waksman, *Microbial Antagonisms and Antibiotic Substances* (New York: Commonwealth Fund, 1945 and 1947 eds.); and Selman Waksman, ed., *Streptomycin: Nature and Practical Applications* (Baltimore: Williams & Wilkins, 1949).

202 **"What are you trying to prove"** Russell Watson to Selman Waksman, February 10, 1954, SAW, box 14, 6.

203 **"This culture was found"** Selman Waksman, *My Life with the Microbes* (New York: Simon and Schuster, 1954), 281.

203 **"fingers of my hand"** Waksman, *My Life*, 203.

204 **"To name only a few"** Waksman, *My Life*, 219.

204 **parable of the sick chicken** Selman A. Waksman, *The Conquest of Tuberculosis* (Berkeley and Los Angeles: University of California Press, 1964), 115–18.

23. The Copied Notebooks

205 **recorded on world-history Web sites** http://www.historyorb.com.

205 **"He Turned Down Millions"** A. E. Hotchner, "He Turned Down Millions," *This Week*, May 30, 1954, 11.

206 **"unique in the discovery"** George Griffenhagen to Selman Waksman, April 16, 1953, SA, Record Unit 7091, Series 5.

207 **"How about notebook pages, etc.?"** Selman Waksman to George Griffenhagen, April 17, 1953, SA, Record Unit 7091, series 5.

207 **"original notes books"** George Griffenhagen to Selman Waksman, April 20, 1953, SA, Record Unit 7091, series 5.

207 **"comprise my various notes"** Selman Waksman to George Griffenhagen, April 23, 1953, SA, Record Unit 7091, series 5.

207 "the most significant notebook" George Griffenhagen to Selman Waksman, April 28, 1953, SA, Record Unit 7091, series 5.

208 "re-copied these experiments" Selman Waksman to George Griffenhagen, May 5, 1953, SA, Record Unit 7091, series 5.

208 a four-page summary Selman Waksman, "Remarks on three of his notebooks containing data on the antagonistic properties of microorganisms and production of antibiotic substances which led to the isolation of streptomycin," May 1, 1953, SA, Record Unit 7091, series 5.

209 "four original experiments" "Historic Hand-written Notes..." Rutgers News Service, July 1, 1953.

209 "Four pages" "Smithsonian Gets Waksman Articles," *New York Times*, July 4, 1953.

Part V: The Restoration

24. Wilderness Years

213 "devotion to science" "Ten Outstanding Young Men," United States Junior Chamber of Commerce awards ceremony, Seattle, January 23, 1954, SAW, box 14, 48.

213 under the headline Maura Devlin, "Schatz, Streptomycin Discoverer, Is Honored," *Bergen Evening Record*, January 14, 1954.

213 "gross exaggeration" Wallace Moreland to United States Junior Chamber of Commerce judges, telegram, January 1953, SAW, box 14, 8.

213 "wide open door" Russell Watson to Selman Waksman, February 10, 1954, box 14, 6.

214 literally "loved to do" Albert Schatz to Milton Wainwright, undated, MW.

214 copper mosses Albert Schatz, "Copper Mosses: Speculations on the Ecology and Photosynthesis of the Copper Mosses," *Bryologist* 58 (June 1955).

214 "offer was gone" Vivian Schatz, author interview, November 8, 2008.

215 "intellectualize" Albert Schatz to Milton Wainwright, undated, 1989, MW.

215 challenged the fertilizer companies Peter Tompkins and Christopher Bird, *Secrets of the Soil* (New York: Harper and Row, 1989), 116.

215 celery farm Vivian Schatz, author interview, November 8, 2008. Also see Albert Schatz to Doris Jones, March 21, 1951, MW.

216 William Wightman, a lecturer in the history *The Growth of Scientific Ideas* (Edinburgh and London: Oliver and Boyd, 1950).

216 "a double act of folly" William Wightman to Dr. J. J. Martin, June 5, 1955, AS.

216 **"no better in these matters"** W. I. B. Beveridge to Dr. J. J. Martin, April 14, 1955, AS personal archive.

216 **"collaborators"** W. I. B. Beveridge, *The Art of Scientific Investigation: An Entirely Fresh Approach to the Intellectual Adventure of Scientific Research* (New York: Vintage Books, 1957), 195.

217 **"had to keep that quiet"** Vivian Schatz, author interview, November 8, 2008. Albert Schatz was elected an Academic Member of the University of Chile, but resigned in 1973 after the assassination of the socialist President Salvador Allende and Chile's takeover in a coup by the military dictatorship of Augusto Pinochet. "My self-respect and sense of human decency compel me to submit my resignation," he wrote. Albert Schatz to Raúl Bazan, December 23, 1973, AS.

218 **blistering attack** Albert Schatz, "Some Personal Reflections on the Discovery of Streptomycin," *Pakistan Dental Review* 15, no. 4 (1965): 125–34.

219 **"unfortunate"** "Great Boon, Sad Story," editorial, *Passaic Herald-News,* November 2, 1965.

219 **eight-page article** S. A. Waksman, "A Quarter Century of the Antibiotic Era," *Antimicrobial Agents and Chemotherapy* (1965): 1–19.

219 **forty-eight "selected" scientific articles** *Scientific Contributions of S. A. Waksman: Selected Articles Published in Honor of His 80th Birthday, July 22, 1968,* ed. H. Boyd Woodruff (New Brunswick, NJ: Rutgers University Press, 1969).

220 **included Waksman's acceptance speech** Boyd Woodruff, author interview, September 21, 2011.

220 **"principal discoverer"** "Selman A. Waksman, Nobel Prizewinner, Dies," *New York Times,* August 17, 1973, 1.

221 **half a day** Boyd Woodruff to Hubert Lechevalier, October 28, 1981, HL.

221 **"strictly a manager of research"** Hubert Lechevalier to Boyd Woodruff, October 19, 1981, HL.

221 **"really interested him"** Hubert Lechevalier to Boyd Woodruff, November 4, 1981, HL.

221 **"systematic development"** Roland Hotchkiss, "Selman Abraham Waksman, July 22, 1888–August 16, 1973," *Biographical Memoirs, The National Academy Press* 83 (2003): 321–39.

221 **"really important discovery"** Bernard Davis, "Two Perspectives on René Dubos, and on Antibiotic Actions," in Carol Moberg and Zanvil Cohn, eds., *Launching the Antibiotic Era: Personal Accounts of the Discovery and Use of the First Antibiotics* (New York: The Rockefeller University Press, 1990), 72.

221 **most concise, comparative** Hubert Lechevalier, "The Search for Antibiotics at Rutgers University," in *The History of Antibiotics: A Symposium*, ed. John Parascandola (Madison, WI: American Institute of the History of Pharmacy, 1980), 116.

25. The English Scientist

223 **young British lecturer** Milton Wainwright, author interview, January 27, 2009.

223 **"major, if largely overlooked scandal"** Milton Wainwright, "Selman A. Waksman and the Streptomycin Controversy," *Society for General Microbiology Quarterly* 15, no. 4 (1988): 90–92.

223 **former student** Hubert Lechevalier to Albert Schatz, December 24, 1992, AS.

224 **"stopped long ago"** Albert Schatz to Doris Jones, July 25, 1983, AS.

224 **recorded their conversation** Albert Schatz interviewed by Milton Wainwright, February 18, 1989.

227 **"make peace with"** Albert Schatz to Milton Wainwright, December 18, 1989, MW.

227 **first popular account** Milton Wainwright, *Miracle Cure: The Story of Penicillin and the Golden Age of Antibiotics* (Oxford, UK: Basil Blackwell, 1990).

227 **"all the more tragic"** Wainwright, *Miracle Cure*, 137.

227 **"met with a blanket"** Milton Wainwright, joint interview with Albert Schatz by Jay Ingram, *Quirks and Quarks*, CBC, October 13, 1990.

26. A Medal

228 **three-page article** Marguerite Smolen, "A Nobel Quest," *Rutgers Magazine*, Winter 1992, 43–45.

228 **Wainwright's two articles** *Society for General Microbiology Quarterly* 15, no. 4 (1988) and "Streptomycin: Discovery and Resultant Controversy," *History and Philosophy of the Life Sciences* 13 (1991): 97–124.

228 **a new book** Frank Ryan, *Tuberculosis: The Greatest Story Never Told* (Bromsgrove, UK: Swift, 1992), 209–23.

228 **three letters** Albert Schatz, letter to the editor, *Rutgers Magazine*, February 4, 5, and 6, 1993, AS.

229 **"errors and omissions"** Lori Chambers to Albert Schatz, February 23, 1992, AS.

229 *Smithsonian* Ken Chowder, "How TB Survived Its Own Death to Confront Us," *Smithsonian*, November 1992, 180–94.

229 **"complete distortion"** Albert Schatz to Don Moser, editor, *Smithsonian*, December 10, 1992 (not published), AS.

229 **Doris Jones also wrote** Doris Jones, letter to the editor, *Smithsonian*, January 12, 1993, AS.

229 **in a letter** Douglas Eveleigh, letter to editor, *Smithsonian*, January 1993.

230 **"At first he wouldn't go"** Vivian Schatz, author interview, November 8, 2008.

230 **"Let's get this over with"** Vivian Schatz, author interview, November 8, 2008.

230 **flyer for the lecture** Biotechnology Club of Cook College, flyer, April 22, 1993, AS.

230 **"It's amazing"** Albert Schatz, "A Lifetime of Research" transcript, April 22, 1993, Rutgers University, New Brunswick, NJ, AS personal archive.

230 **"We got a kick out of that"** Vivian Schatz, author interview, November 8, 2008.

230 **"worldwide impact"** Francis Lawrence, Rutgers Medal ceremony, Rutgers University, New Brunswick, NJ, April 28, 1994.

230 **one local headline** "Overlooked Pioneer Finally Gets His Due," *North Jersey Herald & News*, April 29, 1994.

230 **receives overdue honors** Jeannine DeFoe, *The Daily Targum*, April 29, 1994.

231 **"an intentional grab for glory"** Kitta McPherson, "Rutgers Will Honor a Snubbed Drug Pioneer," *Newark Star-Ledger*, April 25, 1994.

231 *Sciences* Karl Maramorosch, letter to the editor, *Sciences*, January/February 1994.

231 **"cruel law"** Mark Ernest and John Sbarboro, letter to the editor, *The Sciences*, January/February 1994.

231 **his personal view** Byron Waksman, letter to the editor, *The Sciences*, May/June 1994.

232 **if his own contribution had been included** Albert Schatz to Smithsonian Institution, June 28, 1973, SA, Record Unit 613, box 320, 10.

232 **list of Waksman's items** S. Dillon Ripley to Albert Schatz, October 6, 1975, SA, Record Unit 613.

234 **"absolutely no doubt"** Milton Wainwright to Smithsonian, copy, February 20, 1991, MW.

234 **"thoroughly enjoyed"** Patricia Gossel, curator to Albert Schatz, July 26, 1996, AS personal archive.

Afterword

235 **threaten efforts** M. D. Iseman, "Tuberculosis Therapy: Past, Present and Future," *European Respiratory Journal* 20, no. 36 (2002): 87s–94s.

235 **[microbes] communicate** Paul Williams, Klaus Winzer, Weng Chang and Miguel Camara, "Look Who's Talking: Communication and Quorum Sensing in the Bacterial World," *Philosophical Transactions of the Royal Society* 362 (2007): 1119–1134. See also Grace Yim, Helena Huimi Wang, and Julian Davies, "Antibiotics as Signaling Molecules," *Philosophical Transactions of the Royal Society* 362 (2007): 1195–1200.

236 **artificial environments** Julian Davies, "Where Have All the Antibiotics Gone?" *Canadian Journal of Infectious Diseases and Medical Microbiology* 17, no. 5 (2006): 287–90.

236 **"sleeping genes"** David Hopwood, "An Introduction to the Actinobacteria," *Microbiology Today* (May 2007): 60–62.

Index

Note: page numbers in *italics* refer to illustrations.

A Note on the Author

Peter Pringle is the author and coauthor of ten previous books on science and politics, including *Food Inc.*, the bestselling *Those Are Real Bullets: Bloody Sunday, Derry, 1972*, and a mystery-thriller about food and patents, *Day of the Dandelion*. For thirty years, he was a foreign correspondent working for British newspapers, including the *Sunday Times*, the *Observer*, and the *Independent*, in Africa, the Middle East, Europe, the former Soviet Union, and the United States. He has also written for the *New York Times*, the *Washington Post*, the *Atlantic*, and the *New Republic*. He is a graduate of Oxford University and a Fellow of the Linnean Society of London. He lives in New York City and the Adirondacks.